Use R!

Advisors:
Robert Gentleman · Kurt Hornik · Giovanni Parmigiani

Use R!

Series Editors: Robert Gentleman, Kurt Hornik, and Giovanni Parmigiani

Florian Hahne · Wolfgang Huber
Robert Gentleman · Seth Falcon

Bioconductor Case Studies

 Springer

Florian Hahne
Fred Hutchinson Cancer Research Center
Division of Public Health Sciences
Program in Computational Biology
1100 Fairview Avenue, N., M2-B876,
PO Box 19024, Seattle, WA 98109-1024
USA
fhahne@fhcrc.org

Wolfgang Huber
Wellcome Trust Genome Campus
European Bioinformatics Institute
EMBL Outstation Hinxton
Hinxton
Cambridge CB10 1SD
UK
huber@ebi.ac.uk

Robert Gentleman
Program in Computational Biology
Division of Public Health Sciences
Fred Hutchinson Cancer Research Center
1100 Fairview Avenue, N., M2-B876
PO Box 19024, Seattle, Washington 98102-1024
USA
rgentleman@fhcrc.org

Seth Falcon
seth@userprimary.net
http://userprimary.net/user

Series Editors
Robert Gentleman
Program in Computational Biology
Division of Public Health Sciences
Fred Hutchinson Cancer Research Center
1100 Fairview Avenue, N., M2-B876
PO Box 19024, Seattle, Washington 98102-1024
USA

Kurt Hornik
Department für Statistik und Mathematik
Wirtschaftsuniversität Wien Augasse 2-6
A-1090 Wien
Austria

Giovanni Parmigiani
The Sidney Kimmel Comprehensive Cancer Center
at Johns Hopkins University
550 North Broadway
Baltimore, MD 21205-2011
USA

ISBN: 978-0-387-77239-4   e-ISBN: 978-0-387-77240-0
DOI: 10.1007/978-0-387-77240-0

Library of Congress Control Number: 2008926129

Preface

Recent developments in genomics and molecular biology finally carry the promise of understanding the functions of complex biological systems on a whole genome level. These developments have led to enormous amounts of data generated in highthroughput technologies, most prominently in gene expression microarrays. Within the Bioconductor project, an increasing number of researchers are trying to establish solutions for the analysis of such data, combining knowledge from such diverse disciplines as statistics, computer science, bioinformatics, and molecular biology.

With microarrays becoming a standard technology in many molecular biology labs, there is increased demand for comprehensive yet easy to follow instructions to the complex data analysis process. After many years of teaching introductory Bioconductor courses we can identify the main topics of interest, the common misunderstandings and pitfalls, and have learned to better understand key problems with which beginners to the analysis tasks are often challenged. In this book, we try to guide the readers through each step of the data analysis process, beginning from import and data processing to the generation of lists of differentially expressed genes and finally the modeling and interpretation of these lists in downstream analyses. Every chapter focuses on real data use cases that illustrate the problem, and we present both executable code and detailed background information for each step. A companion Webpage to this book can be found at http://www.bioconductor.org/pub/docs/BioconductorCaseStudies

Acknowledgments

We would like to thank Stefano Iacus for organizing the annual Bioconductor courses in Bressanone, Italy, which are the basis for this book and May Alipao, who has helped to organize the Bioconductor courses at the Hutchinson Center. We also thank the many students who attended these and other courses and whose countless questions, helpful remarks, and enthusiasm provided a valuable source of inspiration during the genesis of this book, immensely shaping the outcome you hold in your hands right now. We thank all the co-authors of the individual chapters. Their expert knowledge was highly appreciated and helped to clarify the key concepts and to point out critical steps. We would also like to thank the following individuals who contributed in one or another way to this book: James W. MacDonald, Marc Carlson, Nolwenn LeMeur, Brig Mecham, Joern Toedling, Steffen Durinck, Anna Freni Sterrantino, Deepayan Sarkar, Rafael Irizarry, Jean (Zhijin) Wu, and Li Long.

RG expresses his thanks and appreciation to Tanja and Sophie, for their encouragement and understanding during the long hours spent working on this and other projects.

Florian Hahne
Wolfgang Huber
Robert Gentleman
Seth Falcon

Contents

List of Contributors

V.J. Carey, Channing Laboratory, Brigham and Women's Hospital, Harvard Medical School, Boston, MA, USA

T. Chiang, European Molecular Biology Laboratory, European Bioinformatics Institute, Cambridge, UK

S. Falcon, Scientific Software Engineer

F. Hahne, Computational Biology Program, Fred Hutchinson Cancer Research Center, Seattle, WA, USA

W. Huber, European Molecular Biology Laboratory, European Bioinformatics Institute, Cambridge, UK

M. Morgan, Computational Biology Program, Fred Hutchinson Cancer Research Center, Seattle, WA, USA

D. Scholtens, Department of Preventive Medicine, Northwestern University, Chicago, IL, USA

A. von Heydebreck, Global Technologies, Merck KGaA, Darmstadt, FRG

1

The ALL Dataset

F. Hahne and R. Gentleman

Abstract

In this initial chapter we briefly describe the typical data preprocessing steps for a sample dataset that will be used in many of the
following exercises.

1.1 Introduction

In the course of this book we frequently need a dataset that can be used to
demonstrate the usage of Bioconductor software. For many of the methods
we consider, the initial steps of the analyis procedure are more or less identical. Usually, they comprise subsetting of the data followed by a nonspecific
filtering step to remove probe sets that are not likely to be informative.
In this introductory chapter we briefly guide you through these first steps
needed to obtain data suitable for the various analyses.

1.2 The ALL data

The ALL data consist of microarrays from 128 different individuals with
acute lymphoblastic leukemia (ALL). There are 95 samples with B-cell ALL
and 33 with T-cell ALL and because these are different tissues and quite
different diseases we consider them separately, and typically focus on the
B-cell ALL tumors. Two different analyses have been reported in (Chiaretti
et al., 2004, 2005), which can be consulted for more detail. A number of
covariates are stored along with data, describing more general properties of
the patients such as age or sex but also a whole range of clinical parameters
about type and stage of the disease. Details about these can be found on
the manual page for this dataset. The data have been jointly normalized
using rma and stored in the form of an *ExpressionSet* (see Chapter 2 for
details on this class).

F. Hahne et al., *Bioconductor Case Studies*, DOI: 10.1007/978-0-387-77240-0_1,
© Springer Science+Business Media, LLC 2008

1.3   Data subsetting

We first load the **ALL** package and attach the data to our work space.

```
> library("Biobase")
> library("ALL")
> library("genefilter")
> data("ALL")
```

An interesting subset, with two groups having approximately the same number of samples in each group, is the comparison of the B-cell tumors found to carry the BCR/ABL mutation to those B-cell tumors with no observed cytogenetic abnormalities. These samples are labeled BCR/ABL and NEG in the `mol.biol` covariate. The BCR/ABL mutation, also known as the *Philadelphia chromosome*, was the first cytogenetic aberration that could be associated with the development of cancer, leading the way to the current understanding of the disease. In tumors harboring the BCR/ABL translocation a short piece of chromosome 22 is exchanged with a segment of chromosome 9. As a consequence, a constitutively active fusion protein is transcribed which acts as a potent mitogene, leading to uncontrolled cell division.

Not all leukemia tumors carry the Philadelphia chromosome; there are other mutations that can be responsible for neoplastic alterations of blood cells, for instance a translocation between chromosomes 4 and 11 (ALL1/AF4), and in one of the exercises we also use data from a subset of these tumors.

First, we select those samples originating from B-cell tumors by searching the BT variable (which distinguishes the B-cell from the T-cell tumors) for entries beginning with the letter *B* using a regular expression.

```
> bcell = grep("^B", as.character(ALL$BT))
```

Next, we want to know which of the samples are of molecular types BCR/ABL or NEG.

```
> types = c("NEG", "BCR/ABL")
> moltyp = which(as.character(ALL$mol.biol) %in% types)
```

Combining these two selection criteria gives us a data set for carrying out comparisons between tumors harboring the BCR/ABL translocation and those that did not have any of the tested molecular abnormalities.

```
> ALL_bcrneg = ALL[, intersect(bcell, moltyp)]
```

One more step is required. Some of the sample annotation data is kept in variables of type *factor*. These are variables that can only take a number of discrete categorical values. The set of possible values of a factor is called its levels (you may consult the manual page by typing "? factor"). Because we have reduced the set of samples, we only need fewer factor levels than were present in ALL, and the most succint way to reduce the set of levels of a factor to those that are actually present is by calling the constructor function factor on it again.

```
> ALL_bcrneg$mol.biol = factor(ALL_bcrneg$mol.biol)
> ALL_bcrneg$BT = factor(ALL_bcrneg$BT)
```

1.4   Nonspecific filtering

Some fraction of genes were not expressed at all in the cells that were assayed, at least not to a level that we could detect with the microarrays used here. For further genes, the data did not show enough variation to allow any reliable detection of differential expression. It is a good idea to remove probe sets for these genes before further analysis, and we can do that on the basis of variance. You should consult Chapter 6 and Chapter 7 for more details on nonspecific filtering. Here, we show how to proceed using the function nsFilter from the **genefilter** package to filter for a number of different criteria, all controlled by the function's various parameters. Setting feature.exclude="^AFFX" removes the control probes, which can be identified by the prefix AFFX in their name. As a measure of dispersion for the variance filtering step, we use the interquartile range (IQR), and we choose the 0.5 quantile of the IQR values as a cutoff. This appears to be a reasonable value for the biological setting and the microarray design used here, but it is likely that you may need to adjust this value to your experiment. In Chapter 6 we give an example how to derive a more data-driven cutoff value.

```
> varCut = 0.5
> filt_bcrneg = nsFilter(ALL_bcrneg, require.entrez=TRUE,
      require.GOBP=TRUE, remove.dupEntrez=TRUE,
      var.func=IQR, var.cutoff=varCut,
      feature.exclude="^AFFX")
> filt_bcrneg$filter.log
$numDupsRemoved
[1] 968

$numLowVar
[1] 5212

$feature.exclude
[1] 19
```

```
$numNoGO.BP
[1] 1779

$numRemoved.ENTREZID
[1] 404
> ALLfilt_bcrneg = filt_bcrneg$eset
```

1.5   BCR/ABL ALL1/AF4 subset

There are also other subsets of the data in which we might be interested.
The following code produces a subset consisting of samples from BCR/ABL
positive tumors harboring the t9;22 translocation and ALL1/AF4 positive
tumors with t4;11 translocations.

```
> types = c("ALL1/AF4", "BCR/ABL")
> moltyp = which(ALL$mol.biol %in% types)
> ALL_af4bcr = ALL[, intersect(bcell, moltyp)]
> ALL_af4bcr$mol.biol = factor(ALL_af4bcr$mol.biol)
> ALL_af4bcr$BT = factor(ALL_af4bcr$BT)
> filt_af4bcr = nsFilter(ALL_af4bcr,require.entrez=TRUE,
        require.GOBP=TRUE, remove.dupEntrez=TRUE,
        var.func=IQR, var.cutoff=varCut)
> ALLfilt_af4bcr = filt_af4bcr$eset
```

   Note that nsFilter's default choice of variance measure is not really
appropriate for this case because the sizes of the sample groups differ quite
a lot. See Chapter 8 for a more thourough discussion.

2

R and Bioconductor Introduction

R. Gentleman, F. Hahne, S. Falcon, and M. Morgan

Abstract

In this chapter we cover basic uses of R and begin working with Bioconductor datasets and tools. Topics covered include simple R programming, R graphics, and working with *environments* as hash tables. We introduce the *ExpressionSet* class as an example for a basic Bioconductor structure used for holding genomic data, in this case expression microarray data. And we explore some visualization techniques for gene expression data to get a feeling for R's extensive graphical capabilities.

2.1 Finding help in R

To get started with R and Bioconductor it is important to know where you can find help for the numerous functions, classes, and concepts you are about to come across. The ? operator is the most immediate source of information about R objects. Preceding the name of a function with ? quickly gets you to the manual page of this function. Possible arguments and return values should be introduced there, and you will find basic information about the purpose and application of the function. A special flavor of ? exists for classes. `class ? foo` will get you to the manual page of class `foo` where you will often also find information about available methods for this class.

Function `apropos` can be used to find objects in the search path partially matching the given character string. `find` also locates objects, yet in a more restrictive manner.

F. Hahne et al., *Bioconductor Case Studies*, DOI: 10.1007/978-0-387-77240-0_2,

```
> apropos("mean")
 [1] "colMeans"         "kmeans"
 [3] "mean"             "mean.Date"
 [5] "mean.POSIXct"     "mean.POSIXlt"
 [7] "mean.data.frame"  "mean.default"
 [9] "mean.difftime"    "rowMeans"
[11] "weighted.mean"
> find("mean")
[1] "package:base"
```

If you want to get information about a certain topic or concept, try
help.search. The function searches the help system for documentation
matching a given character string in the (file) name, alias, title, concept, or
keyword entries. Names and titles of the matched help entries are displayed.

```
> help.search("mean")
```

Moreover, there is a wealth of information just waiting for you out
on the Web: A very good introduction is R-Foundation (2007). For
many of the usual R-related questions you may most likely find an
answer in the R-FAQ at http://cran.r-project.org/faqs.html. More
specialized sources for help are the R and Bioconductor mailing lists
(http://www.r-project.org/mail.html, http://www.bioconductor.
org/mailList.html). You can subscribe to different sublists, regarding
your interests and level of expertise and post your questions to the R
society. Before doing so, you should read the posting guides. Often questions on the mailing lists are not answered because major posting rules
have been violated. It is also a good idea to search the online mailing
archives before posting a question. A lot of them have already been asked
and answered by someone else. A searchable Bioconductor archive can be
found at http://dir.gmane.org/gmane.science.biology.informatics.
conductor and the R archives at http://dir.gmane.org/index.php?
prefix=gmane.comp.lang.r.. All of these links can also be found on the
Bioconductor and R-Project Web pages.

Most of the Bioconductor packages contain another valuable source of
information through their package vignettes. Vignettes are supposed to
describe more thoroughly the steps needed to perform one or several of
the specific tasks for which the package was designed. Text and executable
code are bundled together in one document, similar to the document you
are reading right now, which makes it easy to reproduce individual steps on
your own machine or to introduce modifications specific to your own task.
The function openVignette in **Biobase** can be used to open PDF versions of
the available vignettes. Vignettes become available once you load a package
(see the next section for an introduction to the concept of packages).

Exercise 2.1

   a. *There are a number of different plotting functions available. Can you find them?*

   b. *Try to find out which function to use in order to perform a Mann–Whitney test.*

   c. *Open the PDF version of the vignette "Bioconductor Overview" which is part of the **Biobase** package.*

2.2 Working with packages

The design of R and Bioconductor is modular. A lot of the functionality is provided by additional units of software called packages. There are many hundreds of packages available for R and around 260 for Bioconductor. Before we begin working with data, it is important that you learn how to find, download, and install packages.

Different methods can be used for this task, and over time we expect them to become more standardized. R packages are stored in libraries. You can have multiple libraries on your computer, although most people have only one on their personal machine. To add a package to your library, you need to download and install it. After that, each time you want to use the package, you need to load it. You do this using either the `library` function or the function `require`.

Downloading packages can be done using the menu on a distribution of R that has a GUI (Windows or Mac OS X). On these platforms you simply select the packages you want from a list, and they are downloaded and installed. Installing a package does not automatically load it into your R session, you must do that. By default this mechanism will download the appropriate binary packages.

You can use the function `install.packages` to download a specified list of packages. One of the arguments to `install.packages` controls whether package dependencies should also be downloaded and for Bioconductor packages we strongly recommend setting this to `TRUE`.

To make the installation of Bioconductor packages as easy as possible, we provide a Web-accessible script called `biocLite` that you can use to install any Bioconductor package along with its dependencies. You can also use `biocLite` to install packages hosted on CRAN. Here is a sample session illustrating how to use `biocLite` to install the **graph** and **xtable** packages.

```
> source("http://bioconductor.org/biocLite.R")
> biocLite(c("graph", "xtable"))
```

The command `update.packages` can be used to check for and install new versions of already installed packages. Note that you need to supply update.packages with the URL to a Bioconductor repository in order to update Bioconductor packages as well. The recommended way of updating all your installed packages is:

```
> source("http://bioconductor.org/biocLite.R")
> update.packages(repos=biocinstallRepos(), ask=FALSE)
```

Exercise 2.2
What is the output of function sessionInfo?

2.3   Some basic R

Before we begin, let's make sure you are familiar with the basic data structures in R and the fundamental operations that are necessary for both application of existing software and for writing your own short scripts. If you find it easy to answer the following five questions you are ready to proceed with this chapter and learn about the great stuff you can do with your genomic data. If not, it might be a good idea to go back to the excellent "Introduction to R" which you can find on the R Foundation home page at http://cran.r-project.org/manuals/R-intro.html to acquire a more solid foundation of the nitty-gritty details of the language.

Exercise 2.3
  a. *The simplest data structure in R is a* vector. *Can you create the following vectors?*
     – *x with elements* 0.1, 1.1, 2.5, *and* 10
     – *An integer vector y with elements 1 to 100*
     – *A logical vector z indicating the elements of y that are below 10*
     – *A named character vector* pets *with elements* dog, cat, *and* bird. *You can choose whatever names you like for your new virtual pets.*

  b. *What happens to vectors in arithmetic expressions? What is the result of*

     ```
     2 * x + c(1,2)
     ```

  c. *Index vectors can be used to select subsets of elements of a vector. What are the three different types of index vectors? How do we index a matrix or an array?*

  d. *How can we select elements of a list? How do we create a* list?

  e. *What is the difference between a* data.frame *and a* matrix?

2.3.1   Functions

Writing functions in R is easy. All functions take inputs and they return
values. In R the value returned by a function is either specified explicitly by
a call to the function return or it is simply the value of the last expression.
So, the two functions below will return identical values.

```
> sq1 = function(x) return(x*x)
> sq2 = function(x) x*x
```

These functions are *vectorized*. This means you can pass a vector x to the
function and each element of x will be squared. Note that if you use two
vectors of unequal length for any vectorized operation R will try to recycle
the shorter one. Although this can be useful for certain applications, it can
also lead to unexpected results.

Exercise 2.4
*In this exercise we want you to write a function that we use in the next
section. It relies on the R function paste, and you may want to read the
function's manual page. The function should take a string as input and
return that string with a caret prepended. Let's call it ppc; what we want
is that* ppc("xx") *returns* "^xx".

One of the places that user-defined functions are often used is with the
apply family of functions and in the next section we show some examples.

2.3.2   The apply family of functions

In R a great deal of work is done by applying some function to all elements of
a list, matrix, or array. There are several functions available for you to use;
apply, lapply, sapply are the most commonly used. The function eapply is
also available for applying a function to each element of an *environment*.
We show more about how to create and how to work with environments in
the next section.

To understand how the apply family of functions works, we use
them to explore some of the metadata for the Affymetrix®HG-U95Av2
GeneChip®. Because these data are stored in *environments* we make use
of the eapply function.

The hgu95av2MAP *environment* contains the mappings between Affymetrix identifiers and chromosome band locations. For example, in the code
below we find the chromosome band to which the gene, for probe 1001_at
(TIE1), maps.

```
> library("hgu95av2.db")
> hgu95av2MAP$"1001_at"
[1] "1p34-p33"
```

We can extract all of the map locations for a particular chromosome or part of a chromosome by using regular expressions and the `apply` family of functions. First let's be more explicit about the problem: say we want to find all genes that map to the p arm of chromosome 17. Then we know that their map positions will all start with the characters 17p. This is a simple regular expression, ^17p, where the caret, ^, means that we should match the start of the word. We do this in two steps: first we use `eapply` and `grep` and ask for `grep` to return the value that matched.

```
> myPos = eapply(hgu95av2MAP, function(x) grep("^17p", x,
        value=TRUE))
> myPos = unlist(myPos)
> length(myPos)
[1] 190
```

Here we used an anonymous function to process each element of the hgu95av2MAP *environment*. We could have named it and then used it.

```
> f17p = function(x) grep("^17p", x, value=TRUE)
> myPos2 = eapply(hgu95av2MAP, f17p)
> myPos2 = unlist(myPos2)
> identical(myPos, myPos2)
[1] TRUE
```

Exercise 2.5
Use the function ppc that you wrote in the previous exercise to create a new function that can find and return the probes that map to any chromosome (just prepend the caret to the chromosome number) or the chromosome number with a p or a q after it.

2.3.3   Environments

In R, an *environment* is a set of symbol–value pairs. These are similar to lists, but there is no natural ordering of the values and so you cannot make use of numeric indices. Also unlike lists, partial matching of the symbols will not work. Otherwise they behave the same way. In the previous section you have already used an *environment* that stored the mapping between Affymetrix identifiers and chromosome band locations. Here, we show how to work with your own *environment*s.

We first create an *environment* and carry out some simple tasks, such as storing things in it, removing things from it, and listing the contents.

```
> e1 = new.env(hash=TRUE)
> e1$a = rnorm(10)
> e1$b = runif(20)
> ls(e1)
[1] "a" "b"
> xx = as.list(e1)
> names(xx)
[1] "a" "b"
> rm(a, envir=e1)
```

Exercise 2.6

    a. *Create an environment and put in the chromosomal locations of all genes on chromosome 18 using your function from the last exercise.*

    b. *Put into the environment a second function that takes the strings of chromosomal locations and strips the "18" from each string. The function* gsub *can help you with that.*

    c. *Now write a function,* myExtract, *that takes an environment as an argument and returns a vector of stripped chromosomal locations (i.e., apply the stripping function in the environment to the vector of chromosomal locations in the same environment).*

2.4   Structures for genomic data

Genomic data can be very complex, usually consisting of a number of different bits and pieces. In Bioconductor we have taken the approach that these pieces should be stored in a single structure to easily manage the data. The package **Biobase** contains standardized data structures to represent genomic data. The *ExpressionSet* class is designed to combine several different sources of information into a single convenient structure. An *ExpressionSet* can be manipulated (e.g., subsetted, copied), and is the input to or output of many Bioconductor functions.

    The data in an *ExpressionSet* consist of

- assayData: Expression data from microarray experiments (assayData is used to hint at the methods used to access different data components, as we show below).

- metadata: A description of the samples in the experiment (phenoData), metadata about the features on the chip or technology used for the experiment (featureData), and further annotations for the features, for example gene annotations from biomedical databases (annotation).

- experimentData: A flexible structure to describe the experiment.

The *ExpressionSet* class coordinates all of these data, so that you do not usually have to worry about the details. However, an *ExpressionSet* needs to be created in the first place, because it will be the starting point for many of the analyses using Bioconductor software.

In this section we learn how to create and manipulate *ExpressionSet* objects, and by doing that we again practice some basic R skills.

2.4.1   Building an ExpressionSet from .CEL and other files

Many users have access to .CEL or other files produced by microarray chip manufacturer hardware. Usually the strategy is to use a Bioconductor package such as **affyPLM**, **affy**, **oligo**, **limma**, or **arrayMagic** to read these files. These Bioconductor packages have functions (e.g., `ReadAffy`, `expresso`, or `justRMA` in **affy**) to read CEL files and perform preliminary preprocessing, and to represent the resulting data as an *ExpressionSet* or other type of object. Suppose the result from reading and preprocessing CEL or other files is named `object`, and `object` is different from *ExpressionSet*; a good bet is to try, for example,

```
> library(convert)
> as(object, "ExpressionSet")
```

It might be the case that no converter is available. The path then is to extract relevant data from `object` and use this to create an *ExpressionSet* using the instructions below.

2.4.2   Building an ExpressionSet from scratch

As mentioned before, the data from many high-throughput genomic experiments, such as microarray experiments, usually consist of several conceptually distinct parts: assay data, sample annotations, feature annotations, and an overall description of the experiment. We construct each of these components, and then assemble them into an *ExpressionSet*.

Assay data

One important part of the experiment is a matrix of "expression" values. The values are usually derived from microarrays of one sort or another, perhaps after initial processing by manufacturer software or Bioconductor packages. The matrix has F rows and S columns, where F is the number of features on the chip and S is the number of samples.

A likely scenario is that your assay data are in a "tab-delimited" text file (as exported from a spreadsheet, for instance) with rows corresponding to features and columns to samples. The strategy is to read this file into R using the `read.table` command, converting the result to a *matrix*. A typical command to read a tab-delimited file that includes column headers is

```
> dataDirectory = system.file("extdata", package="Biobase")
> exprsFile = file.path(dataDirectory, "exprsData.txt")
> exprs = as.matrix(read.table(exprsFile, header=TRUE,
      sep="\t", row.names=1, as.is=TRUE))
```

The first two lines create a file path pointing to where the assay data are stored; replace these with a character string pointing to your own file, for example,

```
> exprsFile = "c:/path/to/exprsData.txt"
```

(Windows users: note the use of / rather than \; this is because R treats the \ character as an "escape" sequence to change the meaning of the subsequent character.) See the help pages for read.table for more detail. A common variant is that the character separating columns is a comma ("comma-separated values", or "csv" files), in which case the sep argument might be sep=",".

It is always important to verify that the data you have read match your expectations. At a minimum, check the class and dimensions of geneData and take a peek at the first several rows.

```
> class(exprs)
[1] "matrix"
> dim(exprs)
[1] 500   26
> colnames(exprs)
 [1] "A" "B" "C" "D" "E" "F" "G" "H" "I" "J" "K" "L" "M"
[14] "N" "O" "P" "Q" "R" "S" "T" "U" "V" "W" "X" "Y" "Z"
> head(exprs)
                  A      B      C      D      E     F
AFFX-MurIL2_at  192.7  85.75 176.8 135.6 64.49 76.4
AFFX-MurIL10_at  97.1 126.20  77.9  93.4 24.40 85.5
AFFX-MurIL4_at   45.8   8.83  33.1  28.7  5.94 28.3
AFFX-MurFAS_at   22.5   3.60  14.7  12.3 36.87 11.3
AFFX-BioB-5_at   96.8  30.44  46.1  70.9 56.17 42.7
AFFX-BioB-M_at   89.1  25.85  57.2  70.0 49.58 26.1
                  G     H     I      J      K     L      M
AFFX-MurIL2_at  160.5 66.0 56.9 135.61 63.44 78.2 83.1
AFFX-MurIL10_at  98.9 81.7 97.8  90.48 70.57 94.5 75.3
AFFX-MurIL4_at   31.0 14.8 14.2  34.49 20.35 14.2 20.6
AFFX-MurFAS_at   23.0 16.2 12.0   4.55  8.52 27.3 10.2
AFFX-BioB-5_at   86.5 30.8 19.7  46.35 39.13 41.8 80.2
AFFX-BioB-M_at   75.0 42.3 41.1  91.53 39.91 49.8 63.5
```

	N	O	P	Q	R	S	T
AFFX-MurIL2_at	89.3	91.1	95.9	179.8	152.5	180.83	85.4
AFFX-MurIL10_at	68.6	87.4	84.5	87.7	108.0	134.26	91.4
AFFX-MurIL4_at	15.9	20.2	27.8	32.8	33.5	19.82	20.4
AFFX-MurFAS_at	20.2	15.8	14.3	15.9	14.7	-7.92	12.9
AFFX-BioB-5_at	36.5	36.4	35.3	58.6	114.1	93.44	22.5
AFFX-BioB-M_at	24.7	47.5	47.4	58.1	104.1	115.83	58.1

	U	V	W	X	Y	Z
AFFX-MurIL2_at	157.99	146.8	93.9	103.86	64.4	175.62
AFFX-MurIL10_at	-8.69	85.0	79.3	71.66	64.2	78.71
AFFX-MurIL4_at	26.87	31.1	22.3	19.01	12.2	17.38
AFFX-MurFAS_at	11.92	12.8	11.1	7.56	20.0	8.97
AFFX-BioB-5_at	48.65	90.2	42.0	57.57	44.8	61.70
AFFX-BioB-M_at	73.42	64.6	40.3	41.82	46.1	49.41

Sample annotation

The information about the samples (e.g., experimental conditions or parameters, or attributes of the subjects such as sex, age, and diagnosis) is often referred to as covariates. The information describing the samples can be represented as a table with S rows and V columns, where V is the number of covariates. An example of such a table can be input with

```
> pDataFile = file.path(dataDirectory, "pData.txt")
> pData = read.table(pDataFile,
      row.names=1, header=TRUE, sep="\t")
> dim(pData)
[1] 26  3
> rownames(pData)
 [1] "A" "B" "C" "D" "E" "F" "G" "H" "I" "J" "K" "L" "M"
[14] "N" "O" "P" "Q" "R" "S" "T" "U" "V" "W" "X" "Y" "Z"
> summary(pData)
   gender        type          score
 Female:11   Case   :15    Min.   :0.100
 Male  :15   Control:11    1st Qu.:0.328
                           Median :0.415
                           Mean   :0.537
                           3rd Qu.:0.765
                           Max.   :0.980
```

There are three columns of data, and 26 rows. Note that the rows of the sample data table align with the columns of the expression data matrix:

```
> all(rownames(pData) == colnames(exprs))
[1] TRUE
```

This is an essential feature of the relationship between the assay and sample data; *ExpressionSet* will complain if these names do not match.

Sample data can take on a number of different forms. For instance, some covariates might reasonably be represented as numeric values. Other covariates (e.g., gender, tissue type, or cancer status) might be better represented as `factor` objects (see the help page for `factor` for more information). It is important that the sample covariates are encoded correctly; the `colClasses` argument to `read.table` can be helpful in correctly inputting (and ignoring, if desired) columns from the file.

Exercise 2.7

a. *What class does* `read.table` *return?*

b. *Determine the column names of* `pData`. *Hint:* `apropos("name")`.

c. *Use* `sapply` *to determine the classes of each column of* `pData`. *Hint: read the help page for* `sapply`.

d. *What is the sex and Case/Control status of the 15th and 20th samples? What is the status for the sample(s) with* `score` *greater than 0.8?*

Investigators often find that the meaning of simple column names does not provide enough information about the covariate. What is the cryptic name supposed to represent? In what units are the covariates measured? We can create a data frame containing such metadata (or read the information from a file using `read.table`) with

```
> metadata = data.frame(labelDescription=c("Patient gender",
        "Case/control status", "Tumor progress on XYZ scale"),
        row.names=c("gender", "type", "score"))
```

This creates a *data.frame* object with a single column called "labelDescription", and with row names identical to the column names of the *data.frame* containing the sample annotation data. The column `labelDescription` must be present; other columns are optional.

Bioconductor's **Biobase** package provides a class called *AnnotatedDataFrame* that conveniently stores and manipulates tabular data in a coordinated fashion. Create and view an *AnnotatedDataFrame* instance with:

```
> adf = new("AnnotatedDataFrame", data=pData,
        varMetadata=metadata)
> adf
An object of class "AnnotatedDataFrame"
  rowNames: A, B, ..., Z  (26 total)
```

```
varLabels and varMetadata description:
  gender: Patient gender
  type: Case/control status
  score: Tumor progress on XYZ scale
```

Some useful operations on an *AnnotatedDataFrame* include sampleNames, pData (to extract the original pData *data.frame*), and varMetadata. In addition, *AnnotatedDataFrame* objects can be subset much as a *data.frame*:

```
> head(pData(adf))
  gender    type score
A Female Control  0.75
B   Male    Case  0.40
C   Male Control  0.73
D   Male    Case  0.42
E Female    Case  0.93
F   Male Control  0.22
> adf[c("A","Z"), "gender"]
An object of class "AnnotatedDataFrame"
  rowNames: A, Z
  varLabels and varMetadata description:
    gender: Patient gender
> pData(adf[adf$score > 0.8,])
  gender    type score
E Female    Case  0.93
G   Male    Case  0.96
X   Male Control  0.98
Y Female    Case  0.94
```

Annotations and feature data

Metadata on features are as important as metadata on samples, and can be very large and diverse. A single chip design (i.e., collection of features) is likely to be used in many different experiments, and it would be inefficient to repeatedly collect and coordinate the same metadata for each *ExpressionSet* instance. Instead, the idea is to construct specialized metadata packages for each type of chip or instrument. Many of these packages are available from the Bioconductor Web site. These packages contain information such as the gene name, symbol, and chromosomal location. There are other metadata packages that contain the information that is provided by other initiatives such as GO and KEGG. The *annotate* package provides basic data manipulation tools for the metadata packages.

The appropriate way to create annotation data for features is very straight forward: we provide a character string identifying the type of chip used in the experiment. For instance, the data we are using are from the Affymetrix HG-U95Av2 GeneChip :

```
> annotation = "hgu95av2"
```

It is also possible to record information about features that are unique to the experiment (e.g., flagging particularly relevant features). This is done by creating or modifying an AnnotatedDataFrame like that for adf but with row names of the *AnnotatedDataFrame* matching rows of the assay data.

Experiment description

A basic description about the experiment (e.g., the investigator or lab where the experiment was done, an overall title, and other notes) can be recorded by creating a *MIAME* object. One way to create a *MIAME* object is to use the new function:

```
> experimentData = new("MIAME", name="Pierre Fermat",
      lab="Francis Galton Lab",
      contact="pfermat@lab.not.exist",
      title="Smoking-Cancer Experiment",
      abstract="An example ExpressionSet",
      url="www.lab.not.exist",
      other=list(notes="Created from text files"))
```

Usually, new takes as arguments the class name and pairs of names and values corresponding to different slots in the class; consult the help page for *MIAME* for details of available slots.

Assembling an *ExpressionSet*

An *ExpressionSet* object is created by assembling its component parts, and after all this work the final assembly is disappointingly easy:

```
> exampleSet = new("ExpressionSet", exprs=exprs,
      phenoData=adf, experimentData=experimentData,
      annotation="hgu95av2")
```

Note that the names on the right of each equal sign can refer to any object of appropriate class for the argument. See the help page for *ExpressionSet* for more information.

We created a rich data object to coordinate diverse sources of information. Less rich objects can be created by providing less information. A minimal expression set can be created with

```
> minimalSet = new("ExpressionSet", exprs=exprs)
```

Of course this object has no information about sample or feature data or about the chip used for the assay.

2.4.3 ExpressionSet basics

Now that you have an *ExpressionSet* instance, let's explore some of the basic operations. You can get an overview of the structure and available methods for *ExpressionSet* objects by reading the help page:

```
> help("ExpressionSet-class")
```

When you print an *ExpressionSet* object, a brief summary of the contents of the object is displayed (displaying the entire object would fill your screen with numbers):

```
> exampleSet
ExpressionSet (storageMode: lockedEnvironment)
assayData: 500 features, 26 samples
  element names: exprs
phenoData
  sampleNames: A, B, ..., Z  (26 total)
  varLabels and varMetadata description:
    gender: Patient gender
    type: Case/control status
    score: Tumor progress on XYZ scale
featureData
  featureNames: AFFX-MurIL2_at, AFFX-MurIL10_at, ..., 31
  739_at  (500 total)
  fvarLabels and fvarMetadata description: none
experimentData: use 'experimentData(object)'
Annotation: hgu95av2
```

Accessing data elements

A number of accessor functions are available to extract data from an *ExpressionSet* instance. You can access the columns of the sample data (an *AnnotatedDataFrame*) using $:

```
> exampleSet$gender[1:5]
[1] Female Male   Male   Male   Female
Levels: Female Male
> exampleSet$gender[1:5] == "Female"
[1]  TRUE FALSE FALSE FALSE  TRUE
```

You can retrieve the names of the features using `featureNames`. For many microarray datasets, the feature names are the probe set identifiers.

```
> featureNames(exampleSet)[1:5]
[1] "AFFX-MurIL2_at"   "AFFX-MurIL10_at"
[3] "AFFX-MurIL4_at"   "AFFX-MurFAS_at"
[5] "AFFX-BioB-5_at"
```

The unique identifiers of the samples in the dataset are available via the sampleNames method. The varLabels method lists the column names of the sample data:

```
> sampleNames(exampleSet)[1:5]
[1] "A" "B" "C" "D" "E"
> varLabels(exampleSet)
[1] "gender" "type"    "score"
```

Extract the expression matrix and the *AnnotatedDataFrame* of sample information using exprs and phenoData, respectively:

```
> mat = exprs(exampleSet)
> dim(mat)
[1] 500  26
> adf = phenoData(exampleSet)
> adf
An object of class "AnnotatedDataFrame"
  sampleNames: A, B, ..., Z  (26 total)
  varLabels and varMetadata description:
    gender: Patient gender
    type: Case/control status
    score: Tumor progress on XYZ scale
```

Subsetting

Probably the most useful operation to perform on *ExpressionSet* objects is subsetting. Subsetting an *ExpressionSet* is very similar to subsetting the expression matrix that is contained within the *ExpressionSet*: the first argument subsets the features and the second argument subsets the samples. Here are some examples. Create a new *ExpressionSet* consisting of the five features and the first three samples:

```
> vv = exampleSet[1:5, 1:3]
> dim(vv)
Features  Samples
       5        3
> featureNames(vv)
```

```
[1] "AFFX-MurIL2_at"   "AFFX-MurIL10_at"
[3] "AFFX-MurIL4_at"   "AFFX-MurFAS_at"
[5] "AFFX-BioB-5_at"
> sampleNames(vv)
[1] "A" "B" "C"
```

Create a subset consisting of only the male samples:

```
> males = exampleSet[, exampleSet$gender == "Male"]
> males
ExpressionSet (storageMode: lockedEnvironment)
assayData: 500 features, 15 samples
  element names: exprs
phenoData
  sampleNames: B, C, ..., X  (15 total)
  varLabels and varMetadata description:
    gender: Patient gender
    type: Case/control status
    score: Tumor progress on XYZ scale
featureData
  featureNames: AFFX-MurIL2_at, AFFX-MurIL10_at, ..., 31
  739_at  (500 total)
  fvarLabels and fvarMetadata description: none
experimentData: use 'experimentData(object)'
Annotation: hgu95av2
```

2.5   Graphics

Graphics and visualization are important issues when dealing with complex data such as the ones typically found in biological science. In this section we work through some examples that allow us to create general plots in R. Both R and Bioconductor offer a range of functions that generate various graphical representations of our data. For each function there are usually numerous parameters that enable the user to tailor the output to the specific needs. We only touch on some of the issues and tools. Interested readers should look at Chapter 10 of Gentleman et al. (2005a) or for even more detail Murrell (2005).

The function plot can be used to produce dot plots. Read through its documentation (? plot) and also take a look into the documentation for par, which controls most of the parameters for R's base graphics. We now want to use the plot function to compare the gene expression intensities of two samples from our dataset on a log–log scale.

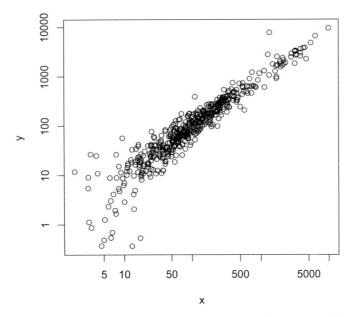

Figure 2.1. Scatterplot of expression intensities for two samples.

```
> x = exprs(exampleSet[, 1])
> y = exprs(exampleSet[, 3])
> plot(x=x, y=y, log="xy")
```

From the plot in Figure 2.1 we can see that the measurements for each probe are highly correlated between the two samples. They form an almost perfect line along the 45 degree diagonal.

Exercise 2.8
The axis annotation of the plot in Figure 2.1 is not very informative. Can you add more meaningful axis labels and a title to the plot? Can you change the plotting symbols? Add the 45 degrees diagonal to the plot. [Hint: use function `abline`.*]*

Proper visualization can help to detect possible problems or inconsistencies in the data. In the simplest case one can spot such problems by looking at distribution summaries. A good example for this is the dependency of the measurement intensity of a microarray probe on its GC-content. To demonstrate this, we need to load a more extended data set from the **CLL** package which includes the raw measurement values for each probe from an experiment using the Affymetrix HG-U95Av2 GeneChip. The `basecontent` function from package **matchprobes** calculates the base frequencies for each probe based on a sequence vector.

```
> library("CLL")
> library("matchprobes")
> library("hgu95av2probe")
> library("hgu95av2cdf")
> library("RColorBrewer")
> data("CLLbatch")
> bases = basecontent(hgu95av2probe$sequence)
```

We now need to match the probes via their position on the array to positions in the data matrix of the CLLbatch object. Because we have several samples in the set, we use the rowMeans function to compute, for each probe, the average of its expression values across arrays.

```
> iab = with(hgu95av2probe, xy2indices(x, y,
      cdf="hgu95av2cdf"))
> probedata = data.frame(
      int=rowMeans(log2(exprs(CLLbatch)[iab, ])),
      gc=bases[, "C"] + bases[, "G"])
```

Now we are ready to plot the $\log_2$-transformed intensity values of the probes grouped by their GC-content. An extraordinarily effective tool for the visualization of distributions is the boxplot. In the code below, we construct a set of boxplots, one box for each possible value of GC-content. The result is shown in Figure 2.2.

```
> colorfunction = colorRampPalette(brewer.pal(9, "GnBu"))
> mycolors = colorfunction(length(unique(probedata$gc)))
> label = expression(log[2]~intensity)
> boxplot(int ~ gc, data=probedata, col=mycolors,
      outline=FALSE, xlab="Number of G and C",
      ylab=label, main="")
```

The first two lines of the above code chunk are concerned with creating a set of colors for the boxes. We use the function brewer.pal from the package **RColorBrewer** to obtain a set of basic colors and colorRampPalette to interpolate between them, defining a unique color for each possible value of gc.

An alternative visualization of univariate distributions is the density plot. We can plot multiple densities in one panel using the function multidensity from the package **geneplotter**. For this plot we focus on the ten biggest groups. First, we find out which of the GC-content values occur most often.

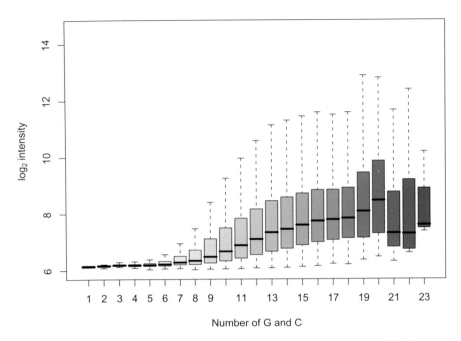

Figure 2.2. Boxplots of the distributions of $\log_2$-intensities from the CLL dataset grouped by GC-content. Please refer to the manual pages of the functions boxplot.stats and boxplot for details on the features shown in a boxplot.

```
> tab = table(probedata$gc)
> gcUse = as.integer(names(sort(tab, decreasing=TRUE)[1:10]))
> gcUse
 [1] 13 11 12 10 14  9 15  8 16 17
```

Then we can call the plot function.

```
> library("geneplotter")
> multidensity(int ~ gc, data=subset(probedata,
      gc %in% gcUse), xlim=c(6, 11),
      col=colorfunction(12)[-(1:2)],
      lwd=2, main="", xlab=label)
```

Exercise 2.9
Another useful distribution summary is plots of the empirical cumulative distribution function. Create a plot similar to the one in Figure 2.3 using the function multiecdf.

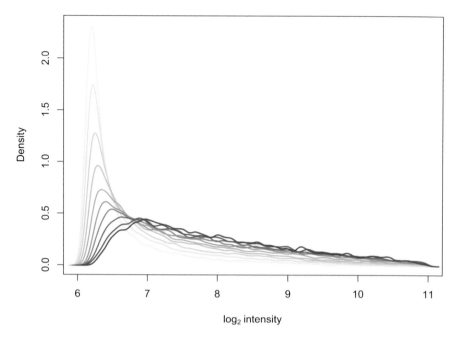

Figure 2.3. Density plot of distributions of $\log_2$-intensities from the CLL dataset grouped by GC content.

3

Processing Affymetrix Expression Data

R. Gentleman and W. Huber

Abstract

In this chapter we do an analysis of Affymetrix gene expression
data. We begin with the CEL files that contain the raw microarray readings, discuss how to do quality assessment, and proceed to
normalization and the estimation of expression values. Finally, we
determine differentially expressed genes.

3.1 The input data: CEL files

The most widely used and generally most useful format in which Affymetrix
data can be obtained is the CEL format. Files in this format are produced
by the array scanner software. This format contains the microarray feature
intensity quantitations, or what might be called "the raw data", and such
data are the starting point for quality assessment and expression analysis.

To import these data into Bioconductor, most users will be served by the
ReadAffy function from the **affy** package.

```
> library("affy")
> myAB = ReadAffy()
```

The function ReadAffy imports CEL files into R objects of class
AffyBatch. These are the basic containers for Affymetrix datasets in Bioconductor. By default, if ReadAffy is called with no further arguments, it
reads all CEL files in the current working directory. The current working
directory can be displayed using the function getwd and be changed with
the function setwd, or on Windows and Mac OS X through the GUI menus.
Alternatively, you can supply the names of the files that you want to import
in the filenames argument, as shown in the next code chunk.

F. Hahne et al., *Bioconductor Case Studies*, DOI: 10.1007/978-0-387-77240-0_3,
© Springer Science+Business Media, LLC 2008

```
> myAB = ReadAffy(filenames=c("a1.cel", "a2.cel", "a3.cel"))
```

In fact, there are different versions of the CEL file format that correspond to different generations of Affymetrix microarrays and of the vendor's software tools. The function ReadAffy is usually able to deal with these automatically, and in most cases you will not need to worry about this. In this chapter we make use of data from a cohort of patients with chronic lymphocytic leukemia (CLL). The data are contained in the **CLL** package. The author of the package has already performed the task of reading the CEL files into an *AffyBatch* object (this saves you time and disk space), so here we can start with that object already. The commands needed to load the object are given next.

```
> library("CLL")
> data("CLLbatch")
> CLLbatch
AffyBatch object
size of arrays=640x640 features (91212 kb)
cdf=HG_U95Av2 (12625 affyids)
number of samples=24
number of genes=12625
annotation=hgu95av2
notes=
```

The data consist of 24 samples run on HG-U95Av2 Affymetrix GeneChip arrays.

3.1.1   The sample annotation

An important aspect of the data analysis paradigm used in the Bioconductor project is that all datasets should be self-contained, self-describing objects. Basically that means that you should be able to find out everything that is relevant about an experiment from the object that is stored and used for the analysis. It also means that a little more effort is needed at the outset to set up the necessary data structures, but this effort is rewarded during all subsequent analyses, when you do not need to worry about manually coordinating expression data, target gene annotation data, sample annotations, and other experimental information. First let us look at the names of the 24 samples in CLLbatch.

```
> sampleNames(CLLbatch)
 [1] "CLL10.CEL" "CLL11.CEL" "CLL12.CEL" "CLL13.CEL"
 [5] "CLL14.CEL" "CLL15.CEL" "CLL16.CEL" "CLL17.CEL"
 [9] "CLL18.CEL" "CLL19.CEL" "CLL1.CEL"  "CLL20.CEL"
[13] "CLL21.CEL" "CLL22.CEL" "CLL23.CEL" "CLL24.CEL"
```

```
[17] "CLL2.CEL"  "CLL3.CEL"  "CLL4.CEL"  "CLL5.CEL"
[21] "CLL6.CEL"  "CLL7.CEL"  "CLL8.CEL"  "CLL9.CEL"
```

As you can see, they are not particularly useful; we need further information about the samples in order to analyze the data.

These data come from 24 patients with CLL. Although a great number of clinical covariates were collected, we restrict our attention to only one, disease status, which is either progressive (abbreviated by *progres.*), stable (*stable*), or missing (*NA*). In practice, such data will often be provided to you in a spreadsheetlike format. In the package **CLL**, they are provided by the *data.frame* object disease.

```
> data("disease")
> head(disease)
  SampleID  Disease
1   CLL10     <NA>
2   CLL11  progres.
3   CLL12    stable
4   CLL13  progres.
5   CLL14  progres.
6   CLL15  progres.
```

As is often the case, we need to do a little data reorganization before starting with the analysis. First, the row names of the *data.frame* should be a suitable set of identifiers, and here we use the SampleID column for this purpose.

```
> rownames(disease) = disease$SampleID
```

Next, we remove the (uninformative) suffixes .CEL from the sample names annotation of CLLbatch, in order to make them match the row names of the disease *data.frame*.

```
> sampleNames(CLLbatch) = sub("\\.CEL$", "",
      sampleNames(CLLbatch))
```

We construct a vector mt that contains the matching of the rows of disease with the samples of CLLbatch.

```
> mt = match(rownames(disease), sampleNames(CLLbatch))
```

Next, we want to provide longer descriptions of the variables so that if we return to the analysis in a few months (or years) we will still be able to see what the variables were. This would also help anyone else who might want to use the data.

```
> vmd = data.frame(labelDescription = c("Sample ID",
       "Disease status: progressive or stable disease"))
```

Finally, we are ready to construct an *AnnotatedDataFrame* object which we can insert into CLLbatch for the annotation of the samples.

```
> phenoData(CLLbatch) = new("AnnotatedDataFrame",
       data = disease[mt, ], varMetadata = vmd)
```

We drop one array for which we have no information on the disease status, because for the analysis we intend this array will not be informative.

```
> CLLbatch = CLLbatch[, !is.na(CLLbatch$Disease)]
```

When you encounter sample annotation data in a spreadsheet, editing into a suitable format will let you use the read.AnnotatedDataFrame function. The description of the format that the function expects and an example file are provided in the **Biobase** package.

3.2 Quality assessment

Quality assessment and quality control (QA/QC) are an essential part of any data analysis. We use the term quality assessment for the computation and interpretation of metrics that are intended to measure quality, and the term quality control for possible subsequent actions, such as removing data from bad arrays or redoing parts of an experiment.

It is a good idea to identify and remove bad arrays as early in the process as you can. There are now several different packages available for QA and we recommend the package **arrayQualityMetrics** and the affyQAReport function from the package **affyQCReport** for generating comprehensive reports. Both of these packages make extensive use of other packages, and both provide easy to read summaries that help to identify arrays that appear to be problematic.

Here, we consider some of the components from which these packages are constructed. The **simpleaffy** package computes a variety of statistics, based primarily on recommendations from Affymetrix, that are intended to assess the quality of the arrays. These statistics are computed using the qc function and can be plotted to obtain a visual overview.

```
> library("affyQCReport")
> saqc = qc(CLLbatch)
> plot(saqc)
```

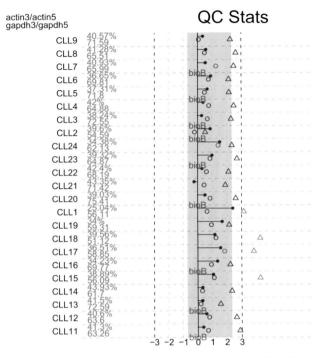

Figure 3.1. QC summary statistics, as produced by the **simpleaffy** package. We see that a number of arrays have actin3/actin5 ratios that are higher than the recommended threshold.

The plot is shown in Figure 3.1.

We can assess whether one or more arrays are different from the others using the `dist2` function on either the raw intensity data, or on normalized expression data. For each pair of arrays, this function computes the median of the absolute values of their differences. This may be considered a measure of distance.

```
> dd = dist2(log2(exprs(CLLbatch)))
```

A heatmap plot of `dd` is created using the code below, and the plot is rendered in Figure 3.2. Because the `levelplot` function does not have a means of automatically reordering rows and columns we do that manually prior to the creation of the plot.

```
> diag(dd) = 0
> dd.row <- as.dendrogram(hclust(as.dist(dd)))
> row.ord <- order.dendrogram(dd.row)
> library("latticeExtra")
> legend = list(top=list(fun=dendrogramGrob,
```

```
    args=list(x=dd.row, side="top")))
> lp = levelplot(dd[row.ord, row.ord],
    scales=list(x=list(rot=90)), xlab="",
    ylab="", legend=legend)
```

The **affyPLM** package (Brettschneider et al., 2007; Bolstad et al., 2005) provides another set of diagnostics that can be used to help assess array quality. There is extensive documentation in that package to help interpret and understand the different approaches that are taken there.

First we need to fit the basic probe-level model.

```
> library("affyPLM")
> dataPLM = fitPLM(CLLbatch)
```

There are two plots that should be examined, the plot of normalized unscaled standard error (NUSE) and that of relative log expression (RLE). In the NUSE plot (Figure 3.3, top), low-quality arrays are those that are significantly elevated or more spread out, relative to the other arrays.

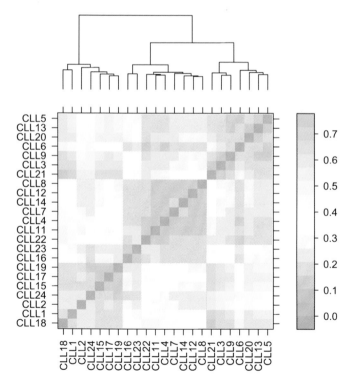

Figure 3.2. Between-array distances, as measured by their median absolute difference.

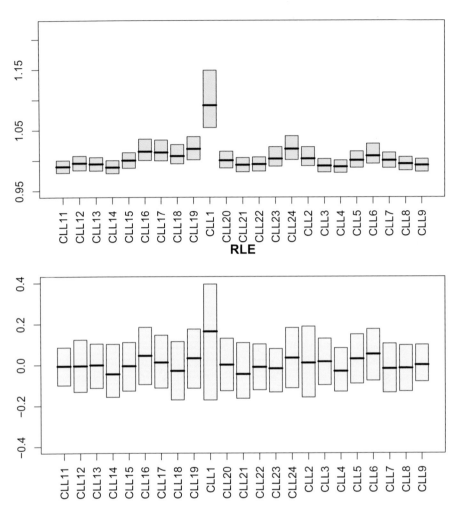

Figure 3.3. Top: NUSE plot; bottom: RLE plot. Both indicate that array CLL1 is potentially problematic.

NUSE values are useful for comparing arrays within one dataset, but their magnitudes are not comparable across different datasets.

```
> boxplot(dataPLM, main="NUSE", ylim = c(0.95, 1.22),
       outline = FALSE, col="lightblue", las=3,
       whisklty=0, staplelty=0)
```

In the RLE plot (Figure 3.3, bottom), problematic arrays are indicated by larger spread, by a center location different from $y = 0$, or both.

```
> Mbox(dataPLM, main="RLE", ylim = c(-0.4, 0.4),
       outline = FALSE, col="mistyrose", las=3,
       whisklty=0, staplelty=0)
```

Both plots in Figure 3.3 indicate that array CLL1 is problematic. Therefore, we drop it from our further analysis.

```
> badArray = match("CLL1", sampleNames(CLLbatch))
> CLLB = CLLbatch[, -badArray]
```

Exercise 3.1
Repeat the calculation of the NUSE and RLE plots for the data with the array CLL1 removed.

3.3 Preprocessing

Expression microarray preprocessing comprises three major tasks: background correction, which aims to adjust the intensity readings for nonspecific signal and hence to increase the array's sensitivity; between-array normalization, which aims to adjust the intensity readings for technical variability between arrays due to subtle differences in handling, labeling, hybridization, and scanning; and reporter summarization, which computes a summary gene expression value for each gene from all the features on the array that target its transcripts. For example, on the HG-U95Av2 GeneChip array there are 409,600 probes that are intended to target 12,625 different gene transcripts.

Between-array normalization methods have been designed for sets of arrays from a single experiment, that is, for arrays which were hybridized under similar conditions with samples that are related and were treated consistently. These methods should not be used to combine microarrays from different experiments, and will often give bad results if used for that purpose. Instead, when there are several separate datasets that are to be analyzed, we recommend preprocessing each set separately, in as similar a fashion as possible, and then using an appropriate statistical model to estimate the effects of interest from the combined data (Gentleman et al., 2005b). To obtain useful estimates biological factors should not be confounded with the experiments.

A comprehensive method that provides all three of the preprocessing tasks is RMA. It is provided by the function rma in the **affy** package (Irizarry et al., 2003).

```
> CLLrma = rma(CLLB)
```

It accepts an instance of the *AffyBatch* class and returns an *Expression-Set* object that can be used in downstream analyses.

The expression values calculated by `rma` are in $\log_2$ scale. A matrix with the estimate expression values can be obtained by using the `exprs` method. The following code extracts the expression data and obtains the dimensions of the matrix containing the data. The same information can also be obtained by calling the `dim` function directly on the `ExpressionSet`.

```
> e = exprs(CLLrma)
> dim(e)
[1] 12625    22
> dim(CLLrma)
Features  Samples
   12625       22
```

In Section 3.5, we explore different aspects of preprocessing in more detail. Furthermore, the `threestep` function in the **affyPLM** package provides a flexible interface for experimenting with different alternative solutions for each of the three tasks, if that is of interest to you.

Exercise 3.2
How many probe sets are there in this dataset?

The sample information that was stored with the raw data `CLLB` has been transferred to the *ExpressionSet*.

```
> pData(CLLrma)[1:3,]
        SampleID  Disease
CLL11     CLL11  progres.
CLL12     CLL12    stable
CLL13     CLL13  progres.
```

You can conveniently use $ to access the sample annotation variables.

```
> table(CLLrma$Disease)
progres.    stable
      14         8
```

3.4   Ranking and filtering probe sets

At this stage of the analysis, we have obtained expression estimates for each gene and each sample. The sensitivity of microarrays is such that we do not expect to reliably detect expression, and differential expression, for more than, say, 50% of the genes on a genome-wide array such as the HG-U95Av2. The signal for the remaining genes will essentially consist of noise and will only aggravate the multiple testing problem (to which we come back in Section 3.4.4). This suggests that some form of nonspecific filtering of noninformative probe sets will have potentially great benefits in the downstream analyses.

The `nsFilter` function in the **genefilter** package can be used to filter out probe sets on a variety of criteria. The one we use here is based on variability. Using the default settings we also filter out probe sets with no Entrez Gene identifiers (they will not map to the annotation data and hence will often not be of much use in downstream analyses) and Affymetrix control probes (their names start with the letters AFFX).

```
> CLLf = nsFilter(CLLrma, remove.dupEntrez=FALSE,
      var.cutof =0.5)$eset
```

3.4.1 Summary statistics and tests for ranking

Log fold-change

A naive first choice for comparing two groups is the average log fold-change. It can be computed by using R base functions such as `rowMeans` applied to the appropriate columns of the matrix `exprs(CLLf)`. We can also use the more convenient function `rowttests`.

```
> CLLtt = rowttests(CLLf, "Disease")
> names(CLLtt)
[1] "statistic" "dm"        "p.value"
```

The function computes for each row the average log-ratio between the two disease groups, the t-statistic, and the corresponding p-value by using the Student's t-distribution.

The variability of log-ratios often depends on the overall intensity of the probe set in question. We can estimate the overall intensity by the average log expression, which is computed in the code chunk below.

```
> a = rowMeans(exprs(CLLf))
```

Exercise 3.3

Plot the log-ratio, `CLLtt$dm`, against the average intensity, a, as in Figure 3.4, to see if the distribution of log-ratios is a function of the average intensity in these data. Also plot log-ratio versus `rank(a)`. Does the variability of the log-ratio values depend on a?

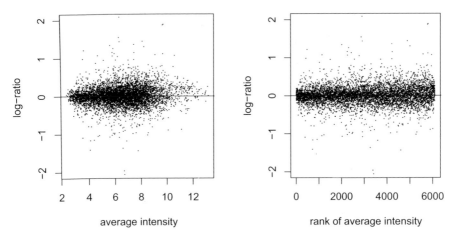

Figure 3.4. Scatterplot of log-ratio versus mean intensity (left panel) and versus rank of mean intensity (right panel).

t-statistic

The *t*-statistic measures the difference in mean divided by an estimate of the variance. When there are few replicates, the variance is not well estimated and the *t*-statistic can perform poorly. Alternative statistics that borrow information across genes often provide better results. An instance of such a modified *t*-statistic is based on an empirical Bayes moderation approach, implemented in the eBayes function in the **limma** package.

```
> library("limma")
> design = model.matrix(~CLLf$Disease)
> CLLlim = lmFit(CLLf, design)
> CLLeb = eBayes(CLLlim)
```

When sample sizes are moderate or large, say ten or more in each group, there is generally no advantage (but also no disadvantage) to using the Bayesian approach.

Exercise 3.4
Compare the t-statistics obtained under the two approaches. Do you agree with the statement above, that for moderate or large sample sizes there is little advantage to the Bayesian approach?

3.4.2   Visualization of differential expression

The volcano plot is a useful way to see the estimate of the log fold-change and statistic you choose to rank the genes simultaneously. Figure 3.5 plots

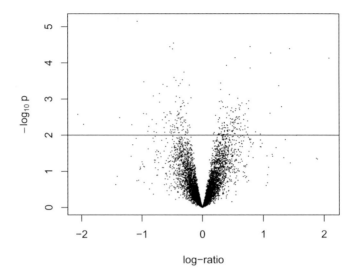

Figure 3.5. A volcano plot using a *t*-statistic: the horizontal line indicates an untransformed *p*-value of 0.01, so points above it have smaller *p*-values than that.

p-values (more specifically, $-\log_{10}$ of the *p*-values) versus effect size. For simplicity we assumed the *t*-statistic follows a *t*-distribution to obtain the *p*-values. It is typical to add some guides to the plot, in the form of either horizontal or vertical lines. In the figures here, we add a horizontal line that corresponds to an untransformed *p*-value of 0.01, so that points above that line have *p*-values less than 0.01.

To create a volcano plot you can use the following code.

```
> lod = -log10(CLLtt$p.value)
> plot(CLLtt$dm, lod, pch=".", xlab="log-ratio",
      ylab=expression(-log[10]~p))
> abline(h=2)
```

Exercise 3.5
Generate a volcano plot using a moderated t-statistic (Figure 3.6). Does it look the same as the plot using the usual t-statistic?

3.4.3   Highlighting interesting genes

Exercise 3.6
Can you highlight the top 25 genes with the smallest p-value (or selected according to another criterion of your choice) with a different color and symbol, as in Figure 3.7? (Hint: look at the function points *with options* col="blue" *and* pch=18.)

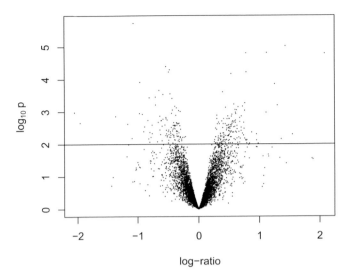

Figure 3.6. A volcano plot when using moderated t-statistic.

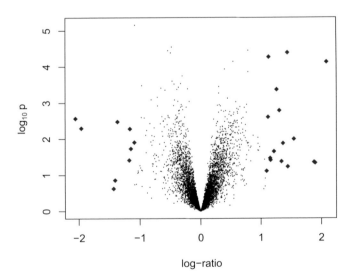

Figure 3.7. A volcano plot with highlighting of the top 25 probe sets.

3.4.4   Selecting hit lists and the multiple testing problem

When you get to this section, you have generated three statistics that can be used to rank genes. Now we turn our attention to deciding on a cutoff.

Suppose we consider all genes attaining p-values less than 0.01. In a single experiment, a p-value of less than 0.01 would be regarded as highly significant. However, here we are testing many hypotheses simultaneously, and the p-values no longer have the conventional meaning. We are testing 6098 null hypotheses (this is the number of probe sets in CLLf) in the example dataset. If they are all true (i.e., if no genes are differentially expressed) we expect $0.01 \times 6098 \approx 61$ hypothesis rejections, which all would be false positives.

Exercise 3.7
*How many probe sets with $p < 0.01$ do you actually see using the t-test?
And with the moderated t-test?*

Various approaches have been suggested for dealing with the multiple testing problem. The **multtest** package provides implementations of many of the options. Alternatively, the function `topTable` in the **limma** package provides multiple testing adjustment methods, including Benjamini and Hochberg's false discovery rate (FDR), simple Bonferroni correction, and several others. For details look at help files for `topTable` and `p.adjust`. The following example lists the top ten genes and creates a report.

```
> tab = topTable(CLLeb, coef=2, adjust.method="BH", n=10)
> genenames = as.character(tab$ID)
```

Here, `coef=2` specifies that we care for the second coefficient in the linear model fit, that is, the between-group difference, as the parameter of interest. The first coefficient is the intercept. The parameter n indicates how many genes should be selected.

3.4.5   Annotation

First we load the **annotate** package,

```
> library("annotate")
```

find out which metadata package we need, and load it.

```
> annotation(CLLf)
[1] "hgu95av2"
> library("hgu95av2.db")
```

Now we can retrieve more annotation information about the genes that interest us, for example, their EntrezGene ID and gene symbol.

```
> ll = getEG(genenames, "hgu95av2")
 1303_at 33791_at 37636_at 36131_at 36939_at 36129_at
  "6452"  "10301"   "9767"   "1192"   "2823"   "9905"
 551_at 39400_at 41776_at 36122_at
  "2033"  "23102"    "475"   "5687"
> sym = getSYMBOL(genenames, "hgu95av2")
  1303_at  33791_at  37636_at   36131_at   36939_at
 "SH3BP2"   "DLEU1"   "PHF16"    "CLIC1"    "GPM6A"
 36129_at     551_at  39400_at   41776_at   36122_at
 "SGSM2"    "EP300"  "TBC1D2B"   "ATOX1"    "PSMA6"
```

We can use the following code to create an HTML page, useful for instance to share results with collaborators.

```
> tab = data.frame(sym, signif(tab[,-1], 3))
> htmlpage(list(ll), othernames=tab,
       filename="GeneList1.html",
       title="HTML report", table.center=TRUE,
       table.head=c("Entrez ID",colnames(tab)))
```

Look for the resulting file *GeneList1.html* in your R working directory. You can open it with your favorite browser, or using the code below.

```
> browseURL("GeneList1.html")
```

To create a report with more annotation information, we can use the **annaffy** package. Below are a few lines of code that create an HTML report with links to various annotation sites.

```
> library("KEGG.db")
> library("hgu95av2.db")
> library("annaffy")
> atab = aafTableAnn(genenames, "hgu95av2.db", aaf.handler())
> saveHTML(atab, file="GeneList2.html")
```

By default `aaf.handler` returns 12 annotation types. You can select a subset instead of reporting all of them.

```
> atab = aafTableAnn(genenames, "hgu95av2.db",
       aaf.handler()[c(2,5,8,12)])
> saveHTML(atab, file="GeneList3.html")
```

3.5   Advanced preprocessing

As noted in Section 3.3, there are three operations that typically need to be performed for microarray data preprocessing: background correction, normalization, and reporter summarization. The "raw" feature intensity data are typically provided in CEL files, one for each scanned array, whereas the assignment of features to target genes is made in the so-called CDF file, of which there is only one for each type of array. In this section, we demonstrate how you can work with these data, in order to construct your own custom analysis methods.

3.5.1   PM and MM probes

Many Affymetrix arrays have both perfect match (PM) and mismatch (MM) probes on the arrays. The PM probes are intended to be complementary to the mRNA being probed, while the MM probes have the same sequence except in the thirteenth position, where the base is changed to its complementary base (e.g. if A in the PM, then T in the MM). The PMs are designed to hybridize specifically to their target cDNA of interest, however, hybridization of short oligonucleotide tends not to be perfectly specific, and there is also a certain amount of nonspecific (or less specific) hybridization from various other cDNA molecules. The intention of the MMs is to measure this nonspecific hybridization. If that were the case, then each MM should generally have lower intensities than its PM partner, and its value could be subtracted from the PM value in order to obtain a more accurate estimation of the signal due to specific hybridization of the target. Let us check this out. The PM and MM values can be extracted using the functions pm and mm.

```
> pms = pm(CLLB)
> mms = mm(CLLB)
```

Exercise 3.8
For the first array in the CLL data, can you make a scatterplot of the PM values versus the MM values, as in Figure 3.8? What do you see, and how might you interpret that? How many MM probes have larger intensities than their corresponding PM probes?

Exercise 3.9
For the second array, can you make a histogram of the MMs for which PM > 2000? As in Figure 3.9, compare it to the histogram where the PM values are less than 2000.

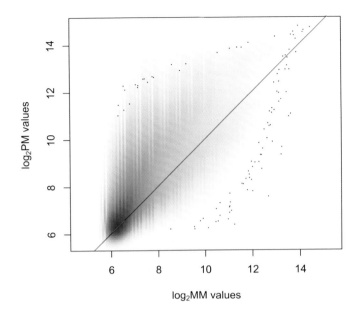

Figure 3.8. A plot of the PM values versus the corresponding MM values for array CLL11.

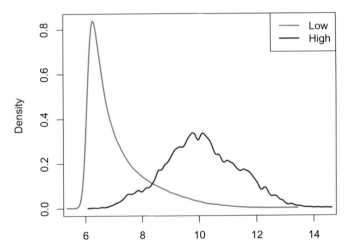

Figure 3.9. Histogram of MM intensities for small (red) and large (blue) values of PM intensities.

3.5.2 Background-correction

The feature intensities from Affymetrix microarrays are always positive, even if the abundance of the intended target gene is zero. This is due to optical noise and, in many cases, cross-hybridization. Background-correction of the intensities is essential to obtain good sensitivity from the data (McGee and Chen, 2006; Ritchie et al., 2007).

In this section, we explore the background-correction method of RMA and compare it to an alternative method. In RMA, background-correction is done by fitting a Normal-Exponential mixture model and subtracting a background estimate from the PM value of each probe that is estimated in such a way that the result is guaranteed to remain positive. Subsequently, the data are logarithm transformed (Irizarry et al., 2003). In VSN, only one overall background estimate is computed for the whole array, and this estimate can be larger than some of the smaller feature intensities on the array. Hence, some of the background-subtracted values can be zero or negative. Subsequently, the so-called generalized logarithm transformation (Huber et al., 2002) is applied, which deals more gracefully with nonpositive values than the logarithm. The RMA background-correction is embedded within the function rma, and it is not easy to get at the background-corrected intensities before the probe set summarization, but we can use the following code to produce them.

```
> bgrma = bg.correct.rma(CLLB)
> exprs(bgrma) = log2(exprs(bgrma))
```

The following code obtains the VSN background correction.

```
> library("vsn")
> bgvsn = justvsn(CLLB)
```

Exercise 3.10
Compare the results of the two background-correction methods to the original values and between each other, as in Figure 3.10. You may want to do the computations on a subset of the data to speed things up.

The return value of the call to justvsn above is an *AffyBatch* with values that have been background-corrected, normalized between arrays, and $\log_2$-transformed. In order to do the summarization of probe sets, we could call the function rma with arguments normalize=FALSE and background=FALSE, or we can use the function vsnrma, which is a wrapper that performs all of these steps.

```
> CLLvsn = vsnrma(CLLB)
```

We can repeat the same analysis as in Section 3.4 to do nonspecific filtering and testing for differential expression.

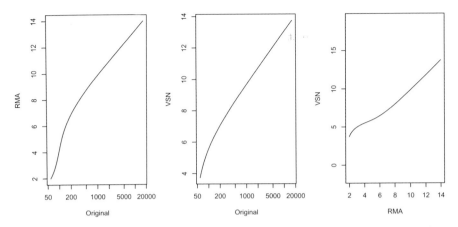

Figure 3.10. Left: the RMA background-correction transformation: the original PM intensities are plotted along the x-axis (on a logarithmic scale), the RMA background-corrected and $\log_2$-transformed values along the y-axis. Without background-correction, the curve would be a straight line. The higher slope of the curve in the low-intensity range reflects the background subtraction. Middle: similar to left panel, but with VSN background-corrected and glog_2-transformed values along the y-axis. Right: Comparison of RMA background-corrected and $\log_2$-transformed values with the VSN background-corrected and glog_2-transformed values.

```
> CLLvsnf = nsFilter(CLLvsn, remove.dupEntrez=FALSE,
      var.cutoff=0.5)$eset
> CLLvsntt = rowttests(CLLvsnf, "Disease")
```

Exercise 3.11
Compare the results with those obtained for CLLtt *in Section 3.4. Produce a scatterplot between the t-statistics obtained for both cases, as in Figure 3.11.*

3.5.3 Summarization

We next show how to explore other probe set summarization methods. First, we create a list of indices. Each element of the list indices corresponds to a probe set and contains the row indices of the matrices pms and mms corresponding to the probe set's probes. These data are obtained from the CDF file.

```
> pns = probeNames(CLLB)
> indices = split(seq(along=pns), pns)
```

As you see, there are indeed 12,625 probe sets.

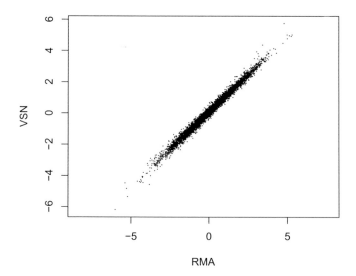

Figure 3.11. Comparison of t-statistics obtained from data preprocessed with RMA using default settings (x-axis) and with VSN for background-correction and between-array normalization (y-axis). For the data shown here, there is practically no difference.

```
> length(indices)
[1] 12625
> indices[["189_s_at"]]
 [1] 15874 15875 15876 15877 15878 15879 15880 15881
 [9] 15882 15883 15884 15885 15886 15887 15888 15889
```

Exercise 3.12
Can you plot the PM and MM intensities for the probes of one probe set across a set of arrays, as in Figure 3.12?

Let us try out a naive summary method: for each sample, we take the median of differences between the PM and MM values for each probe set.

```
> newsummary = t(sapply(indices, function(j)
      rowMedians(t(pms[j,]-mms[j,]))))
> dim(newsummary)
[1] 12625    22
```

The code yielded a matrix with one row for each probe set and one column for each sample.

189_s_at

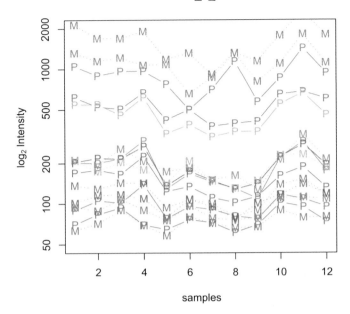

Figure 3.12. Intensities of eight PM probes and their associated MM probes from the probe set `189_s_at`, which targets the PLAUR gene. Shown are the data for the first 12 samples of the CLLB dataset. Note how the profiles of most (but not all) of the probes are correlated across samples, but have very different baselines (this is the so-called *probe effect*). Also, in some cases the MM values are larger than the PM values.

Exercise 3.13

Most biologists are unhappy when expression estimates are negative. What percent of probe sets, for each array, yield negative values for each array?

4

Two-Color Arrays

Florian Hahne and Wolfgang Huber

Abstract

In this case study, two RNA samples are compared to each other on
60 mer oligonucleotide microarrays using two-color labeling. The lab
covers data import, visualization, exploration and normalization of
the data, and the identification of differentially expressed genes.

4.1 Introduction

Transcription-profiling two-color microarrays are hybridized with two samples of cDNA that are obtained from the RNA samples of interest by reverse
transcription. Each of the two samples is labeled with a different dye. A popular choice is the Cy3 and Cy5 dyes. The dyes have different excitation
wavelengths, so the amount of bound cDNA from each of the two samples
at each position on the array can be determined by two passes of an optical scanner. This results in two grey-scale image files which are typically
stored in TIF format. The images are quantified by specialized image analysis software that produces numeric summaries for each feature on the array.
Typically, we are most interested in the overall feature intensity for each
of the two wavelengths. Sometimes, additional statistics are useful, such
as a measured background intensity in the vicinity of the feature, which is
intended to estimate the background signal that would be obtained even
in the absence of the labeled cDNA sample, and spot shape descriptors,
which may be used to detect features of poor manufacturing quality.

The analysis in this lab starts from these numeric feature summaries.
There are many different file formats, and many variations of the technology, so importing the data into a suitable R data structure can be the
first challenge. The function `read.maimages` in the package **limma** provides
excellent functionality for this task; only if your image analysis software
produces an esoteric file format, will you need to adapt the function. In

F. Hahne et al., *Bioconductor Case Studies*, DOI: 10.1007/978-0-387-77240-0_4,

any case, you will need to make some decisions: which of the feature summaries do you want to use for the subsequent statistical analysis, and which identifier system for the features and the reporter sequences on the array?

4.2   Data import

In this lab, we use the *CCl4* example dataset by Holger Laux, Timothy Wilkes, Amy Burrell, and Carole Foy from LGC Ltd. in Teddington, UK. In the experiment, rat hepatocytes were treated either with carbon tetrachloride (CCl_4) or with dimethylsulfoxid (DMSO). In the early twentieth century, CCl_4 was widely used as a dry cleaning solvent, as a refrigerant, and in fire extinguishers, however, it was found to have multiple toxic and possible cancerogenous sideeffects. DMSO is commonly used as a solvent in pharmaceutical applications, and here it served as negative control. Total RNA was hybridized to Agilent® *Rat Whole Genome* microarrays. The arrays use a two-color labeling scheme (Cy3 and Cy5), and the experiment was done as a direct comparison with dye-swaps and three replicates each. The integrity of the RNA was quantified from the electrophoretic trace of the RNA samples by Agilent's RNA Integrity Number (RIN); (Schroeder et al., 2006). The initial samples had a RIN of 9.7. To study the effect of RNA degradation, additional samples were generated by degrading the CCl_4-treated RNA sample with ribonuclease A, resulting in RINs of 5.0 and 2.5. The experimental design is described in more detail below.

Let us load the package **CCl4** that contains these data.

```
> library("limma")
> library("CCl4")
> dataPath = system.file("extdata", package="CCl4")
```

The variable `dataPath` contains the name of the directory in which the input data files are provided.

```
> dir(dataPath)
```

```
[1] "013162_D_SequenceList_20060815.txt"
[2] "251316214319_auto_479-628.gpr"
[3] "251316214320_auto_478-629.gpr"
[4] "... (16 more files) ..."
[5] "samplesInfo.txt"
```

Exercise 4.1
Use a text editor or a spreadsheet program to view these files. What does each of them contain?

Let us have a closer look at the `samplesInfo.txt` file. This is also called the *targets* file. It contains one row for each array in the experiment. Its columns describe the two cDNA samples hybridized to each array; in the current case, there are four columns describing the RNA source and the RNA quality of each of the two RNA samples. Another column is the name of the corresponding image analysis file, which among other information contains the array identifier. In general, the `samplesInfo.txt` file can have columns that refer to properties of the arrays and properties that refer to either of the two cDNA samples hybridized to the array.

We can read this file into R using

```
> adf = read.AnnotatedDataFrame("samplesInfo.txt",
      path=dataPath)
> adf
An object of class "AnnotatedDataFrame"
  rowNames: 251316214319_auto_479-628.gpr, 251316214320_auto_4
  78-629.gpr, ..., 251316214394_auto_463-521.gpr  (18 total)
  varLabels and varMetadata description:
    Cy3:  name of the RNA sample that was labeled with Cy3
    Cy5:  name of the RNA sample that was labeled with Cy5
    RIN.Cy3:  Agilent's RNA integrity number for the Cy3 RNA s
  ample
    RIN.Cy5:  Agilent's RNA integrity number for the Cy5 RNA s
  ample
```

`adf` is an object of class *AnnotatedDataFrame*. This class is very similar to the *data.frame* class that you might know from the R **base** package, but in addition it allows you to annotate the variables that correspond to its columns with more explicit information on the definition of each variable, how it was measured, in which units, and the like. The targets file can be prepared using a text editor or a spreadsheet program such as Microsoft Excel or OpenOffice Calc.

Now let us read the intensity data into an *RGList* object in R. The data files were produced by *Genepix®* image analysis software. We can use the function `read.maimages` for this purpose. This function expects a *data.frame* with mandatory column `FileName` as its input, and we construct this from the object `adf`.

```
> targets = pData(adf)
> targets$FileName = row.names(targets)
```

```
> RG = read.maimages(targets, path=dataPath, source="genepix")
```

Consult the help entry for `read.maimages` to see which other file formats are supported. If the data files contain identifiers of the array features and annotation of their target genes, these are read from the first intensity data file.

```
> head(RG$genes)
  Block Row Column        ID           Name
1     1   1      1 BrightCorner    BrightCorner
2     1   1      2 BrightCorner    BrightCorner
3     1   1      3   (-)3xSLv1 NegativeControl
4     1   1      4 A_44_P317301        AW523361
5     1   1      5 A_44_P386163     NM_001007719
6     1   1      6 A_44_P353916     NM_001025631
```

It is important to note that all data files are assumed to contain data for the same array layout, that is, for the same target genes in the same order. It is your responsibility to confirm that this is true; the `read.maimages` function does not do any checking.

4.3   Image plots

It is interesting to look at the variation of background values over the array. Consider image plots of the red and green background for the first array:

```
> par(mfrow=c(5,1))
> imageplot(log2(RG$Rb[,1]), RG$printer, low="white",
        high="red")
> imageplot(log2(RG$Gb[,1]), RG$printer, low="white",
        high="green")
> imageplot(rank(RG$Rb[,1]), RG$printer, low="white",
        high="red")
> imageplot(rank(RG$Gb[,1]), RG$printer, low="white",
        high="green")
> imageplot(rank(log(RG$R[,1])+log(RG$G[,1])),
        RG$printer, low="white", high="blue")
```

The output of this is shown in Figure 4.1.

4.4   Normalization

An MA-plot displays the log-ratio of red intensities R and green intensities G on the y-axis versus the overall intensity of each spot on the x-axis. The log-ratio is

$$M = \log_2 R - \log_2 G = \log_2(R/G)$$

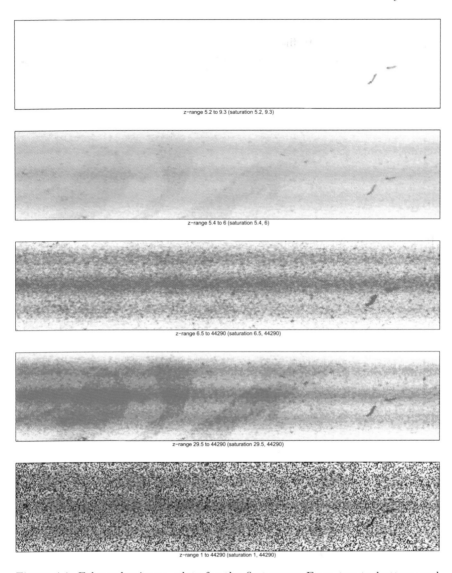

Figure 4.1. False color image plots for the first array. From top to bottom: red background with a logarithmic color scale, green background with a logarithmic color scale, red background with a rank color scale, green background with a rank color scale, rank of the average of logarithms of red and green foreground. In all of the images, we can see traces of a scratch towards the right of the images. Also, there is a pattern of horizontal stripes. In the bottom plot, we can see a gradient from left to right. The color scale determines how the feature intensity values are mapped to the range of colors. Often, the rank scale is more sensitive for the detection of patterns than the logarithmic or untransformed scales. Patterns or scratches such as seen here may affect the downstream analysis, but it is difficult to say from these plots alone how consequential they really are for the bottomline result.

and the overall intensity is measured by

$$A = \frac{1}{2}\left(\log_2 R + \log_2 G\right) = \log_2 \sqrt{RG}.$$

It is easy to compute these using the basic R arithmetic functions; there is also a function in **limma** for this purpose.

```
> MA = normalizeWithinArrays(RG, method="none",
      bc.method="none")
```

With the option `method="none"`, we instruct the function `normalize-WithinArrays` not to do any adjustments to the computed M values. With the option `bc.method="none"`, we specify that only the foreground `R` and `G` values should be used.

We can produce the scatterplot between the M and the A values for the first array using the following instructions.

```
> library("geneplotter")
> smoothScatter(MA$A[, 1], MA$M[, 1], xlab="A", ylab="M")
> abline(h=0, col="red")
```

Exercise 4.2

   a. *What does the plot look like when you use* `bc.method="subtract"` *in the above call to* `normalizeWithinArrays`?

   b. *Why did we use the function* `smoothScatter`? *Try also the* `plotMA` *function from the* **limma** *package. How does it differ?*

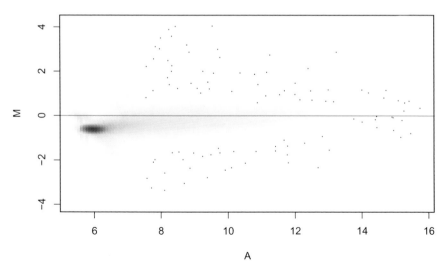

Figure 4.2. MA-plot for the first array. Note the downward curvature of the distribution at the lower end of the A range: for low intensity spots, M tends to be negative, that is, the red color channel is dimmer than the green.

In Figure 4.2, we noted that there is an imbalance between the red and green intensities. There are also differences in the intensity distributions between the different arrays, as we can see in the boxplot that is shown in Figure 4.3.

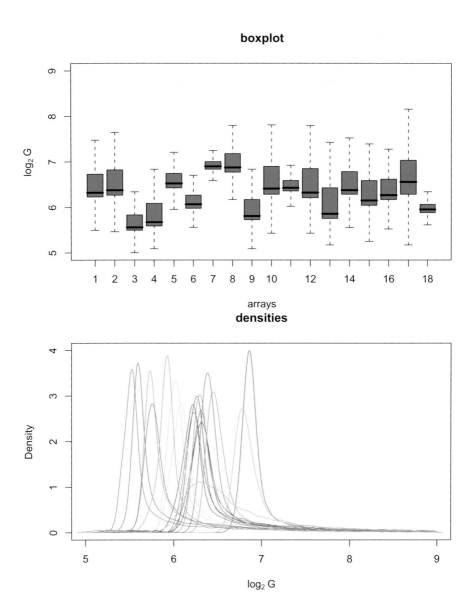

Figure 4.3. Boxplots and smooth density estimates of the distributions of the $\log_2 G$ values of the 18 arrays.

```
> plotformula = log2(RG$G)~col(RG$G)
> boxplot(plotformula, ylim=c(5,9), outline=FALSE,
    col="forestgreen", xlab="arrays",
    ylab=expression(log[2]~G), main="boxplot")
```

In addition to the boxplot, we can also look at the distributions of the G values on the different arrays with the `multidensity` function.

```
> multidensity(plotformula, xlim=c(5,9),
    main="densities", xlab=expression(log[2]~G))
```

The result is shown in the lower panel of Figure 4.3.

The purpose of normalization is to adjust the data for unwanted imbalances, drifts, or biases, such as those that we have seen above between the color channels and between arrays. To keep things simple, in this lab we focus on the subset of six arrays which were hybridized with the good RNA sample of the CCl_4 treated hepatocytes.

```
> rin = with(MA$targets, ifelse(Cy5=="CCl4", RIN.Cy5,
    RIN.Cy3))
> rin
 [1] 9.7 5.0 2.5 2.5 9.7 5.0 2.5 9.7 9.7 5.0 2.5 9.7 5.0 5.0
[15] 2.5 9.7 5.0 2.5
> select = (rin == max(rin))
> RGgood = RG[, select]
> adfgood = adf[select, ]
```

We have also created a subset version `adfgood` of the annotated dataframe `adf`, which we need later on. Note the different indexing conventions: for objects of type *RGList*, like `RG`, arrays are considered as "columns", whereas for the object `adf`, which is of type *AnnotatedDataFrame*, arrays are considered as "rows". We use the function `justvsn` from the **vsn** package to normalize the data from these 6 arrays. A more detailed introduction to VSN is given in Chapter 5.

```
> library("vsn")
> ccl4 = justvsn(RGgood, backgroundsubtract=TRUE)
```

The resulting object `ccl4` is of class *NChannelSet*. A useful plot for checking whether the **vsn** normalization worked well is the scatterplot of standard deviations versus the rank of the mean intensity of each feature. The result is shown in Figure 4.4.

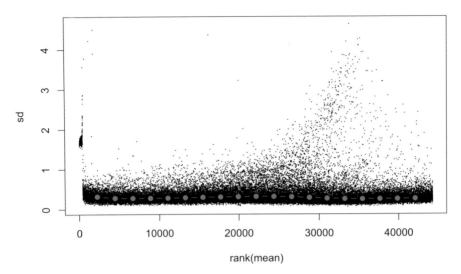

Figure 4.4. For each feature, this plot shows the empirical standard deviation of the normalized and glog-transformed data on the y-axis versus the rank of the mean on the x-axis. The red dots, connected by lines, show the running median of the standard deviation. It should be approximately horizontal, that is, show no substantial trend.

```
> r = assayData(ccl4)$R
> g = assayData(ccl4)$G
> meanSdPlot(cbind(r, g))
```

In the next paragraph, we produce MA-scatterplots of the normalized and glog-transformed data. However, before we proceed, let us first pay some attention to the experimental metadata in the ccl4 object. Because RGgood does not contain all the information that is needed for a valid and complete *NChannelSet*, let us do some additional postprocessing of adfgood at this point. First, we remove the .gpr extension from the sample names, to be consistent with the output of **limma**'s read.maimages function.

```
> rownames(pData(adfgood)) = sub("\\.gpr$", "",
      rownames(pData(adfgood)))
> pData(adfgood)
```

	Cy3	Cy5	RIN.Cy3	RIN.Cy5
251316214319_auto_479-628	DMSO	CCl4	9.0	9.7
251316214330_auto_457-658	CCl4	DMSO	9.7	9.0
251316214333_auto_487-712	DMSO	CCl4	9.0	9.7
251316214379_auto_443-617	CCl4	DMSO	9.7	9.0
251316214382_auto_481-674	DMSO	CCl4	9.0	9.7
251316214391_auto_475-599	CCl4	DMSO	9.7	9.0

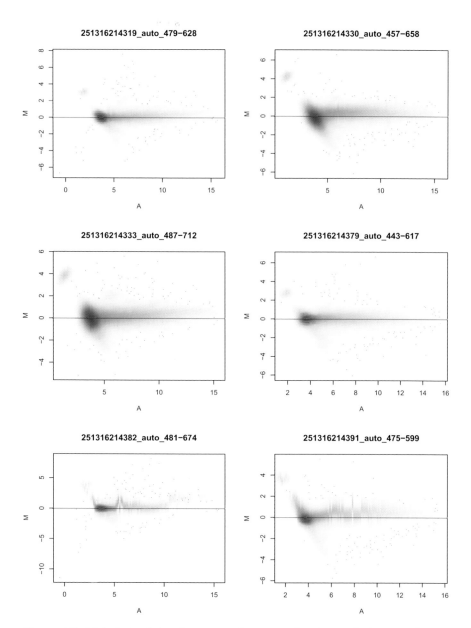

Figure 4.5. MA-plots of the six arrays after normalization and glog-transformation with **vsn**. Can you figure out what causes the little "clouds" above the $M = 0$ line in the left side of the plot?

We also need to update the `channel` information in the `varMetadata` part of `adfgood`. This indicates which of the columns refer to which channel.

```
> varMetadata(adfgood)$channel = factor(c("G", "R", "G", "R"),
     levels = c("G", "R", "_ALL_"))
```

The *factor* variable `channel` needs to have a level `_ALL_`, which is designated for annotation columns that refer to the whole array. In the case of `adfgood`, however, all its four columns are channel-specific. Now we are ready to stick it onto `ccl4`.

```
> phenoData(ccl4) = adfgood
> validObject(ccl4)
[1] TRUE
```

Now let us construct an additional object, `ccl4AM`, that contains the difference (`M=r-g`) and average (`A=(r+g)/2`) glog$_2$ intensities.

```
> ccl4AM = ccl4
> assayData(ccl4AM) = assayDataNew(A=(r+g)/2, M=r-g)
```

Let us also update the `channel` information in the variable metadata of the new object `ccl4AM`: because the two channels `A` and `M` are a combination of both the red and the green intensities, all of the sample annotation variables are now common (`_ALL_`), rather than channel-specific.

```
> varMetadata(phenoData(ccl4AM))$channel[] = "_ALL_"
> validObject(ccl4AM)
[1] TRUE
```

```
> smoothScatter(assayData(ccl4AM)$A[,2],
                assayData(ccl4AM)$M[,2])
> abline(h=0, col="red")
```

The above code produces the MA-plot of the second array. The plots for all six arrays are shown in Figure 4.5.

Here you have seen how to use the function `justvsn` from the package **vsn**. There are also several other alternative normalization methods in the package **limma**. Please see the **limma** vignette and the manual pages for the functions `normalizeWithinArrays` and `normalizeBetweenArrays`.

4.5 Differential expression

Now we are ready to identify the differentially expressed genes in this experiment. First, we need to construct a so-called *model matrix*. This matrix is a general mechanism used in the **limma** package to specify multiway comparisons in complex microarray experimental designs. Here, the design is rather simple: we just consider direct comparisons between two samples, hepatocytes treated with the negative control (DMSO) and with CCl_4.

```
> design = modelMatrix(pData(ccl4AM), ref="DMSO")
Found unique target names:
 CCl4 DMSO
                              CCl4
251316214319_auto_479-628     1
251316214330_auto_457-658    -1
251316214333_auto_487-712     1
251316214379_auto_443-617    -1
251316214382_auto_481-674     1
251316214391_auto_475-599    -1
```

We can now call **limma**'s lmFit function to fit a linear model to the data, separately gene per gene. As the independent variable, the linear model takes the M value for the gene; as the dependent variables, it takes the sample annotations, as provided through the matrix design.

```
> fit = lmFit(assayData(ccl4AM)$M, design)
```

Because of the small number of arrays, the use of the moderated t-statistic is advisable (Tusher et al., 2001; Lönnstedt and Speed, 2002; von Heydebreck et al., 2004; Smyth, 2004). In **limma**, this is done with the eBayes function.

```
> fit = eBayes(fit)
```

The return value of this is an object of class *MArrayLM*.

```
> class(fit)
[1] "MArrayLM"
attr(,"package")
[1] "limma"
> names(fit)
 [1] "coefficients"     "rank"             "assign"
 [4] "qr"               "df.residual"      "sigma"
 [7] "cov.coefficients" "stdev.unscaled"   "pivot"
[10] "genes"            "method"           "design"
[13] "df.prior"         "s2.prior"         "var.prior"
[16] "proportion"       "s2.post"          "t"
[19] "p.value"          "lods"             "F"
[22] "F.p.value"
```

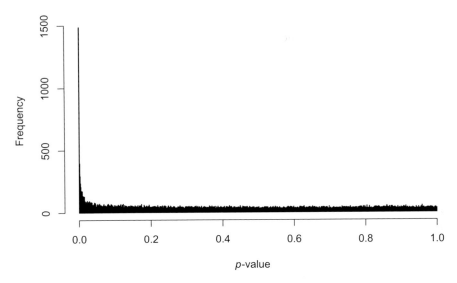

Figure 4.6. Histogram of the raw per-gene p-values from the moderated t-test. The horizontal "floor" of values in this plot corresponds to the large number of features whose target genes are not differentially expressed. The sharp peak at $p < 0.001$ corresponds to differentially expressed genes. The shape of this histogram can be used to assess an experiment and its analysis: if the peak on the left is missing, this indicates lack of power of the experiment or of the analysis to detect differentially expressed genes. If the remainder of the distribution is not fairly uniform, this can indicate overdispersion and/or a strong interfering effect of another variable, for example, an unintended batch effect.

A first sanity check of the result can be obtained by looking at the histogram of the raw per-gene p-values

```
> hist(fit$p.value, 1000)
```

The result of this is shown in Figure 4.6. Visually there seem to be plenty of differentially expressed genes. Let us obtain a summary table of some key statistics for the top ones.

```
> fit$genes = pData(featureData(ccl4AM))
> topTable(fit, number=10, adjust="BH")
      Block Row Column        ID       Name logFC     t
29304     1 285     52 A_44_P312606  BG153336  3.19  33.7
8210      1  80     73 A_44_P312605  BG153336  4.58  33.3
34156     1 332     63 A_44_P548559  XM_342120  3.35  31.1
20886     1 203     80         <NA>      <NA>  3.04  29.5
40623     1 395     41 A_44_P196172  NM_138881  3.49  29.0
18619     1 181     79 A_44_P445344  NM_024134  3.23  27.5
21888     1 213     52 A_44_P139673  NM_031572 -2.49 -27.4
```

21065		1 205		53 A_44_P953483		TC534656	2.40	25.5
24327		1 237		19 A_43_P15719	NM_001012197		2.87	24.9
16195		1 158		24 A_44_P109342		NM_031971	2.47	23.8
	P.Value	adj.P.Val	B
29304	5.26e-10	1.31e-05	11.7
8210	5.91e-10	1.31e-05	11.7
34156	1.02e-09	1.50e-05	11.4
20886	1.53e-09	1.56e-05	11.2
40623	1.76e-09	1.56e-05	11.1
18619	2.69e-09	1.79e-05	10.9
21888	2.83e-09	1.79e-05	10.8
21065	4.98e-09	2.76e-05	10.5
24327	6.01e-09	2.96e-05	10.4
16195	8.67e-09	3.69e-05	10.1

In the table, t is the empirical Bayes moderated t-statistic. The corresponding p-values have been adjusted for multiple testing to control the false discovery rate and B is the empirical Bayes log odds of differential expression (Smyth, 2004). BG153336 is the identifier of an IMAGE clone that has been mapped to the Smarcb1 gene (SWI/SNF related,

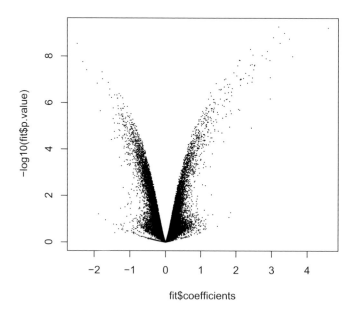

Figure 4.7. The volcano plot shows, for each microarray feature, a measure of effect size (average log-ratio) on the x-axis versus a measure of significance (negative logarithm of the p-value) on the y-axis. Note how features with the same effect size can have quite different significance, and vice versa.

matrix-associated, actin-dependent regulator of chromatin, subfamily b, member 1).

We can look at the volcano plot, which relates p-values (on the y-axis) and effect size (on the x-axis). It is shown in Figure 4.7.

```
> plot(fit$coefficients, -log10(fit$p.value), pch=".")
```

Finally, let us export the table of results of the statistical analysis into a tabulator-delimited file, which you can import, for example, into a spreadsheet program.

```
> write.fit(fit, file="fit.tab")
```

5

Fold-Changes, Log-Ratios, Background Correction, Shrinkage Estimation, and Variance Stabilization

W. Huber

Abstract

Microarray data are affected by experimental variability, which is a combination of systematic and stochastic variability. The basic task of microarray preprocessing is to extract quantities of interest from the data while correcting for systematic variations and controlling the stochastic variability. In this exercise we explore the concepts of (log-)ratios, the role of background correction, the idea of shrinkage estimation, and the generalized logarithm. Some tools for this are provided by the package **vsn**.

5.1 Fold-changes and (log-)ratios

Microarray measurements in most cases carry no meaningful physical units. This is in contrast to many other measurement instruments: clocks measure seconds, weight scales measure kilograms, current meters measure ampères. No such universal units are associated with the feature intensity values measured on a microarray. Although conceptually it is possible to define suitable units (e.g., the number of target RNA molecules per cell), such a calibration is rarely attempted in practice. However, one of the basic tenets of microarray analysis is that expression values for the *reporter* (probe sequence) can be compared across conditions. This motivates why we consider ratios or log-ratios

$$q = \log_2 \frac{I_1}{I_2}, \tag{5.1}$$

F. Hahne et al., *Bioconductor Case Studies*, DOI: 10.1007/978-0-387-77240-0_5,
© Springer Science+Business Media, LLC 2008

where I_1 and I_2 represent the intensities measured for the same reporter across different conditions: either on two different arrays when a one-channel microarray platform was used, or the two different color channels in the case of a two-channel microarray. The use of $\log_2$, the logarithm to base 2, is a popular convention in the field. Underlying the (log-)ratio is the assumption of proportionality

$$I_1 \approx k \cdot c_1, \quad I_2 \approx k \cdot c_2, \tag{5.2}$$

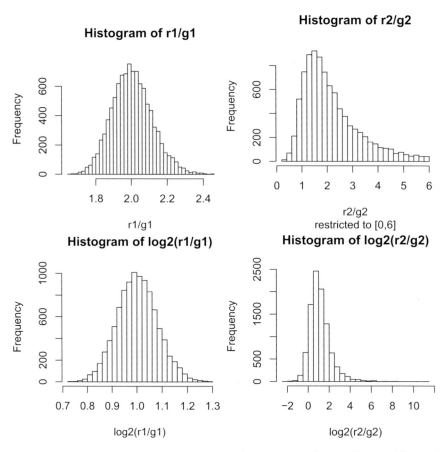

Figure 5.1. Top left: histogram of r1/g1, where r1 was drawn from a Normal distribution with mean 2000 and standard deviation 50, and g1 from one with mean 1000. Top right: histogram of r2/g2, where r2 was drawn from a Normal distribution with mean 200 and standard deviation 50, and g2 from one with mean 100. The histograms in the bottom row show the same data as those in the upper row, but with log-ratios instead of ratios. Their distribution is less asymmetric.

where k is an unkown constant, c_1 is the true abundance of the target molecule in condition 1, measurable in appropriately defined physical units, for example, molecules per cell, and c_2 is the abundance in condition 2. k can be (and usually will be) different for each reporter. By taking the ratio in Equation (5.1), k cancels out and we need not worry about the exact definition of the units of c_1 and c_2. So it seems that the quantity I_1/I_2 is an obvious estimator for the expression fold-change c_1/c_2. In the following exercise, we see that considering the log-ratio has some advantages over the ratio.

Exercise 5.1

a *Generate two random samples* r *and* g *of Normal distributed values, one with a mean of 2000 and standard deviation 50, and one with a mean of 1000 and standard deviation 50. Plot their histogram, as in Figure 5.1.*

b *Now repeat the above, but with two random samples that have means of 200 and 100, respectively (use the same standard deviation as before). Also plot their histogram.*

c *What do the distributions look like when you consider the log-ratio* log2(r/g) *instead?*

5.2   Background-correction and generalized logarithm

Let us load two example datasets. The first is the `kidney` data that come with the **vsn** package,

```
> library("vsn")
> data("kidney")
```

`kidney` is an *ExpressionSet* object with 8704 features and two samples. The samples were obtained from two adjacent locations in kidney tissue from a nephrectomy and their cDNA was labeled with Cy3 and Cy5. The human cDNA microarray was spotted by Holger Sültmann at DKFZ Heidelberg (Sültmann et al., 2005).

The second example dataset is in the **CCl4** package.

```
> library("CCl4")
> data("CCl4")
```

It compares the transcription profiles of rat hepatocytes treated with carbon tetrachloride (CCl_4) and with DMSO as a negative control. Extracted cDNA was labeled with Cy3 and Cy5 and hybridized to Agilent *Rat Whole*

Genome microarrays. Please consult the manual page for more information on these data.

```
> ? CC14
```

Here, we only consider the six arrays for which undegraded ("good") RNA was used, and drop those features that have no ID.

```
> selArrays = with(pData(CC14),
      (Cy3 == "CC14" & RIN.Cy3 > 9) |
      (Cy5 == "CC14" & RIN.Cy5 > 9))
> selFeatures = !is.na(featureData(CC14)$ID)
> CC14s = CC14[selFeatures, selArrays]
```

Exercise 5.2
Plot the histograms of the feature intensities of the two color channels for the kidney *data and for some of the* CC14s *arrays, as in Figure 5.2. (Hint: use the function* multidensity *from the* **geneplotter** *package.)*

We can now contrast the data shown in Figure 5.2 with the predictions of the model (5.2), the model that underlies the use of the log-ratio. The microarray used for the CC14 dataset was genomewide, and because not all genes are expressed in hepatocytes, we expect that a certain fraction of the features on the array targets unexpressed genes. However, all the intensities are well above zero. Clearly, there is a nonzero background signal that is measured even in the absence of the target molecules. Background-correction methods offer a solution, and indeed in Figure 5.2 we see that for the kidney data, which were background-corrected, many of the intensities fall around zero. However, this immediately creates a problem for the log-ratio: the log-ratio does not result in meaningful values for zero or negative intensities. We must conclude that either the background correction method or the concept of taking the log-ratio needs more thought.

Note that simply ignoring the weakly or unexpressed genes will not take us very far: those cases where a gene is off under one condition and gets switched on in another condition may be the most interesting ones, for example, in a disease, as a response to a stimulus, or during a differentiation process.

To better understand what to do, the following extension to the overly simplistic model (5.2) is helpful.

$$I_1 \approx k \cdot c_1 + b_1, \quad I_2 \approx k \cdot c_2 + b_2, \tag{5.3}$$

where b_1, b_2 are positive numbers that represent the background signal. Now what happens to the log-ratio? We get

$$q = \log_2 \frac{I_1}{I_2} \approx \log_2 \frac{kc_1 + b_1}{kc_2 + b_2} \tag{5.4}$$

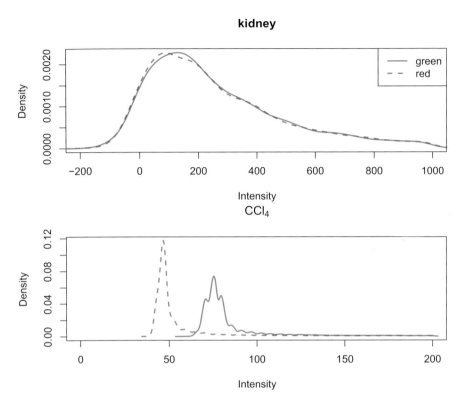

Figure 5.2. Top: density estimates (smoothed histograms) of the green (solid line) and red (dashed line) color channels of the kidney data. The range of the data is $[-561, 35481]$, the x-range of the plot has been limited to $[-200, 1000]$ for better display. The intensities in the `kidney` dataset have been obtained from microarray image analysis software by subtracting a local background estimate from the computed feature intensities. For some of the features, the local background estimate, obtained from pixels surrounding each feature, is higher than the feature quantitation. Often this happens for dim features due to estimation noise. As an effect of the background subtraction, the locations of the two distributions are similar. Bottom: density estimates of the red and green color channels of an array from the CCl4s data. The solid lines correspond to the green channel, the dashed lines to the red. The range of the data is $[37, 65535]$; the x-range of the plot has been limited to $[0, 200]$ for better display. In contrast to the kidney data, no local background subtraction was performed, and all intensities are positive. Also, the locations of the distributions are more disparate. A further point to note is that a large fraction of the data is concentrated at very low intensities; most features are very dim, and only a few have intensities well above background.

and this quantity is compressed towards 0 in comparison to the true logartihmic expression fold-change $\log_2(c_1/c_2)$. Due to the background signal, we are losing sensitivity to detect differential expression. The idea of background-correction is to estimate b_1 and b_2 and to subtract them from the intensities, resulting in $I_1^{bc} = I_1 - \hat{b}_1$ and $I_2^{bc} = I_2 - \hat{b}_2$. The log-ratio of these values can then be used as the fold-change estimate. However, this creates another problem: if the estimator for the b_i is unbiased, then it will sometimes, because of noise in the data, produce estimates that are a bit higher than the true value, and sometimes a bit lower. If at the same time c_i is 0, this means that the background corrected intensities I_i^{bc} can be zero or negative. We have seen this in the kidney example in Figure 5.2. The log-ratio of nonpositive values produces nonsensical results.

Three of the most common solutions to this problem are the following.

First, give up on background-correction and just use the uncorrected values I_i, accepting the resultant loss in sensitivity.

Second, use a biased estimator $\hat{b}_i$ that ensures that I_i^{bc} is always positive (Irizarry et al., 2003; McGee and Chen, 2006; Ritchie et al., 2007).

Third, instead of the log-ratio $\log_2(I_1^{bc}) - \log_2(I_2^{bc})$ use the generalized log-ratio (Huber et al., 2002; Durbin et al., 2002)

$$\text{glog}_2(I_1^{bc}) - \text{glog}_2(I_2^{bc}) \tag{5.5}$$

The generalized logarithm with parameter $a \in \mathbb{R}$ is defined as

$$\text{glog}_2(x) = \log_2 \frac{x + \sqrt{x^2 + a^2}}{2}. \tag{5.6}$$

This is the approach used in the package **vsn**. The theoretical motivation for the choice of the function (5.6) is given in the references (Huber et al., 2002, 2003, 2005). Essentially, the function $\text{glog}_2(x)$ smoothly interpolates between the usual logarithm, $\log_2(x)$, when x is large, and a linear function, $x/a + \log(a/2)$, when x is small. $\text{glog}_2(x)$ is nonsingular and well behaved even for $x \approx 0$, whereas $\log_2(x)$ has a singularity at 0.

Both of the latter two approaches require further specification: in the case of biased background correction, one needs to decide how large that bias should be in each case. With the generalized logarithm, one needs to choose the parameter a. We explore some of this in the following. In practice, the results from using biased background-correction followed by the usual logarithm and from using an unbiased background-correction followed by the generalized logarithm are often surprisingly similar. In Section 5.7, we discuss the interpretation of the glog_2-ratio as a shrinkage estimator of the $\log_2$ fold-change.

Exercise 5.3

    a. *Plot the graph of the generalized logarithm.*

$$\text{glog}_2(x) = \log_2 \left(\frac{1}{2}x + \frac{1}{2}\sqrt{x^2 + a^2} \right) \tag{5.7}$$

$\log_2(a/2) + \dfrac{x}{a \ln(2)}$

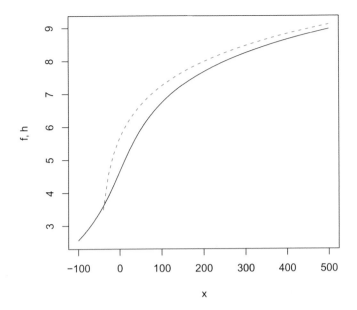

Figure 5.3. Graphs of the functions $f(x) = \log_2(x + b)$ (dashed) and $h(x) = \log_2\left(\frac{1}{2}x + \frac{1}{2}\sqrt{x^2 + a^2}\right)$ (solid) for $a = b = 50$.

Compare it to the graph of the so-called started logarithm $\log_2(x + b)$ (Rocke and Durbin, 2003). You may use $a = b = 50$ and an x-range from -100 to 500, as in Figure 5.3, or try out different values for these parameters.

b. How do the two functions behave as $x \to \infty$?

c. Optional, difficult. How do the parameters a and b correspond to each other so that the started log and the glog behave as similarly as possible when applied to data?

Exercise 5.4

Make, for one of the arrays in the CC14s dataset, a scatterplot of the red versus the green intensities on the untransformed scale. As in Figure 5.4, restrict the axis limits to c(30, 300). Draw the line through the origin $y = x$ into this plot. Also add the line $y = 18 + 1.2x$. Which one seems to fit the data better? How could you formalize this question and find the optimal parameters of the line $y = y_0 + s \cdot x$?

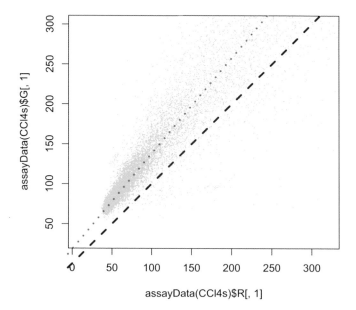

Figure 5.4. For one of the arrays from the CC14s dataset, the scatterplot shows the red versus the green intensities, for values up to 300. The dashed line is the intersect through the origin, $y = x$; the dotted line corresponds to $y = 18 + 1.2x$.

5.3   Calling VSN

Now let us use the function justvsn from the **vsn** package to normalize the CC14s data. Depending on the speed of your computer, this can take a little while, but it should not be more than a few minutes.

```
> CC14sn = justvsn(CC14s, backgroundsubtract=TRUE)
> class(CC14sn)
```

CC14sn is an *NChannelSet* object. The following lines extract the so-called A and M values; essentially these correspond to a 45 degree rotation of the scatterplot of red versus green, such that the average intensity is on the x-axis and the difference on the y-axis.

```
> asd = assayData(CC14sn)
> A = (asd$R+asd$G)/2
> M = (asd$R-asd$G)
```

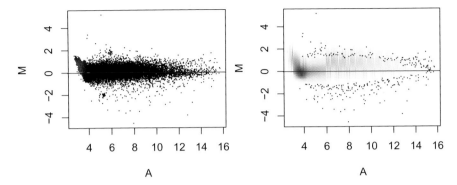

Figure 5.5. MA-plot of the first array in the normalized data CCl4sn. The scatterplot in the left panel uses an individual dot for each data point. The right panel shows a smooth density representation of the data, individual points are drawn only in the sparse regions.

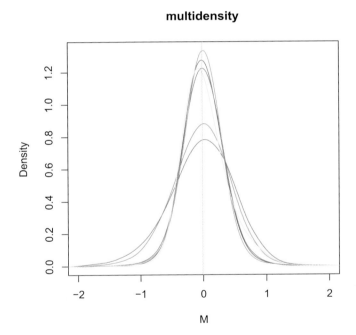

Figure 5.6. Smooth density estimates for the M values of the six arrays in the CCl4sn data. For each array, the distribution is centered at 0, but two of the arrays stand out for having a much wider distribution than the other four. Because all six arrays have been hybridized with the same pair of biological samples, this difference in width of the M distributions may indicate variability in the data quality. Further investigation is necessary to make a more definitive conclusion.

We can produce a normal scatterplot,

```
> plot(A[,6], M[,6], pch='.', asp=1, xlab="A", ylab="M")
> abline(h=0, col="blue")
```

or a variant of it that uses a false color representation of the point density.

```
> smoothScatter(A[,6], M[,6], nrpoints=300, asp=1,
      xlab="A", ylab="M")
> abline(h=0, col="blue")
```

The plots are shown in Figure 5.5.

To assess the normalized data, we can look at the histograms (smooth density estimates) of the M values (Figure 5.6).

```
> multidensity(M, xlim=c(-2,2), bw=0.1)
> abline(v=0, col="grey")
```

5.4   How does VSN work?

VSN tries to find for each array and, if the arrays use more than one dye channel, for each of the dye channels, a background offset and a scaling factor to make the data across arrays as similar as possible. Then it applies the glog-transformation, as discussed in Section 5.2, and the differences of the resulting values, the generalized log-ratios, can be used to quantify differential expression.

Table 5.1. vsn glossary.

vsn	The name of the R package.
vsn2	The basic model fitting function: it accepts *eSet*-like objects (i. e., for the purpose of this glossary, objects of class *ExpressionSet*, *NChannelSet*, *AffyBatch*, *RGList*, and *matrix*) and returns a fit object of class vsn.
predict	A function that accepts a fit object of class vsn and applies it to an *eSet*-like object. It returns an object of the corresponding class.
justvsn	A wrapper around vsn2 and predict that accepts an *eSet*-like object and returns an object of the same class, with the expression data replaced by the normalized values. justvsn(x, ...) is equivalent to fit = vsn2(x, ...); predict(fit, newdata = x, useDataInFit = TRUE).
vsn	The basic model-fitting function in versions 1.x of the package; this function is obsolete now.

Formally, we can write this as follows. The VSN-normalized intensities are, up to an overall additive offset,

$$h_{ij} = \mathrm{glog}_2 \frac{I_{ij} - b_j}{k_j} = m_i + \varepsilon_{ij}, \qquad (5.8)$$

where the index $j = 1, \ldots, d$ counts over the arrays and, if applicable, dye channels, i counts over the array features, b_j is a per-array background-correction term, k_j is a per-array scale factor, m_i is the average intensity, on the glog_2 scale, of feature i, and ε_{ij} are the residuals. VSN tries to find those values for the parameters b_j, k_j, and m_i that minimize the residuals ε_{ij}, and to this end uses an outlier-resistant variant of least squares regression called *least trimmed sum of squares* (LTS) regression (Rousseeuw and Leroy, 1987; Huber et al., 2002). Different values of the parameter a of the glog_2 function in Equation (5.6) can be absorbed by the parameters b_j, k_j, and m_i, so we can just as well set $a = 1$ without loss of generality.

An important tool for assessing whether the VSN fit worked is the plot of m_i versus the empirical standard deviation σ_i. σ_i is computed row by

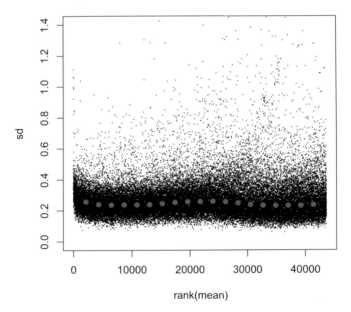

Figure 5.7. Standard deviation versus mean relationship for the CCl4sn data. For each feature, the plot shows the empirical standard deviation σ_i of the normalized and glog-transformed data on the y-axis versus the rank of the mean m_i on the x-axis. The red dots, connected by lines, show the running median of the standard deviatiqon. The rank of the smallest value is 1, the rank of the largest value is 43,628 (the number of features in CCl4sn). The rank scale is used for the x-axis in order to distribute the data evenly along the x-dimension. This allows for a better visual assessment of mean-variance trends compared to when the original scale is used.

row from the matrix of residuals ε_{ij},

$$\sigma_i^2 = \frac{1}{d-1} \sum_{j=1}^{d} \varepsilon_{ij}^2. \tag{5.9}$$

Figure 5.7 shows the result of calling the function meanSdPlot on the 12 × 43628 matrix of red and green intensities in CC14sn.

```
> meanSdPlot(cbind(assayData(CC14sn)$R, assayData(CC14sn)$G),
      ylim=c(0, 1.4))
```

What should such a plot look like? We want the distribution of σ_i to be concentrated at small values, and not see a significant trend of these values as a function of the mean. Features with large σ_i have very different intensities across the arrays (and/or channels); this could either be for a biological reason, because they match a differentially expressed transcript, or it could be a technical artifact that makes the data of certain features excessively variable.

The other fitted parameters of Equation(5.8) can be examined by using the coef method for vsn objects, and interested readers are referred to the package vignette and the manual page of the vsn2 function.

5.5   Robust fitting and the "most genes not differentially expressed" assumption

An important assumption that underlies the algorithm to fit Equation (5.8), and hence the VSN method, is that for all features, except for a minority of outliers, the residuals ε_{ij} should be small. In other words, the assumption is that for many features the abundance of their target is approximately constant across arrays (and/or channels). Features that target differentially expressed transcripts act as outliers.

In the following exercise, we explore the robustness of VSN against such outliers. To this end, let us take the kidney data and computationally "spike in" 33% of the features as if their targets were strongly upregulated.

```
> kidspike = kidney
> sel = runif(nrow(kidspike)) < 1/3
> delta = 100 + 0.4*abs(rowMeans(exprs(kidspike)[sel,]))
> exprs(kidspike)[sel,] = exprs(kidspike)[sel,] +
      cbind(-delta,+delta)
```

The resulting data look somewhat artificial, as we see in Figure 5.8, but are appropriate to investigate our question. Let us call vsn2 on these data, using three choices of the parameter lts.quantile.

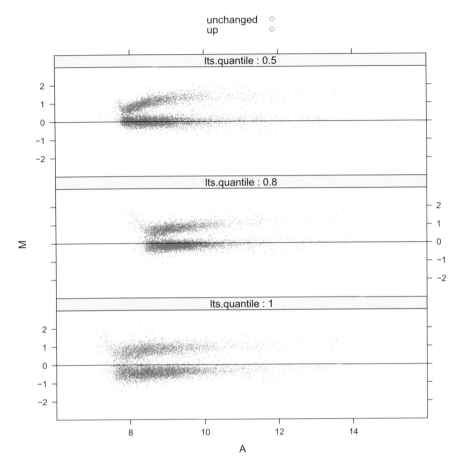

Figure 5.8. Robustness and the "most genes not differentially expressed" assumption. The three panels show results of using VSN on an (artificial) dataset with a high fraction of differentially expressed features. Data for 2919 features, shown in violet, have been computationally "spiked-in" as if their targets were strongly upregulated. The VSN algorithm was not aware of the distinction between the blue and violet data points. For lts.quantile=1, which corresponds to ordinary nonrobust least sum of squares regression, the fit is heavily affected by the violet outliers. For lts.quantile=0.5, which corresponds to least trimmed sum of squares regression with a trimming quantile $q = 50\%$, the blue data points are distributed tightly around the $M = 0$ line, and the algorithm has managed to disregard the outliers. The result for lts.quantile=0.8 is in between. Note how the outliers not only affect the estimation of the array scaling factors (k_j in Equation (5.8)), but also of the background-correction offsets (b_j in Equation (5.8)). This explains why the difference among the three panels is not just a shift in M-direction, but also a change in the shape of the distributions of the transformed data.

```
> ltsq = c(1, 0.8, 0.5)
> vkid = lapply(ltsq, function(p) vsn2(kidspike,
      lts.quantile=p))
```

vkid is a list of three *vsn* objects, one for each choice of lts.quantile. In order to use the **lattice** package to visualize the result, we need to rearrange the data into a suitable *data.frame*. The function getMA takes a *vsn* object and returns a *data.frame* with columns A and M. We apply it to vkid.

```
> getMA = function(x)
      data.frame(A = rowSums(exprs(x))/2,
          M = as.vector(diff(t(exprs(x)))))
> ma = lapply(vkid, getMA)
```

Now we add two further columns to each *data.frame*: group is used for the coloring of the points, and lts.quantile to arrange the data in three panels. Finally, we can join the resulting list of data.frames into a single one using the function rbind.

```
> for(i in seq(along=ma)) {
      ma[[i]]$group = factor(ifelse(sel, "up", "unchanged"))
      ma[[i]]$lts.quantile = factor(ltsq[i])
  }
> ma = do.call("rbind", ma)
```

Now we are ready to plot Figure 5.8.

```
> library("lattice")
> lp = xyplot(M ~ A | lts.quantile, group=group, data=ma,
      layout = c(1,3), pch=".", ylim=c(-3,3), xlim=c(6,16),
      auto.key=TRUE,
      panel = function(...){
          panel.xyplot(...)
          panel.abline(h=0)},
      strip = function(...) strip.default(...,
          strip.names=TRUE, strip.levels=TRUE))
> print(lp)
```

How does the robust *least trimmed sum of squares* (LTS) regression algorithm work (Rousseuw and Leroy, 1987; Huber et al., 2002)? To answer this question, it is instructive to contrast it with ordinary least squares regression. Ordinary least squares tries to minimize the sum of squared residuals, that is, the sum of all ε_{ij}^2 in Equation (5.8). Because of this, one or a few outlier data points that have very large residuals can dominate the fit result. In spite of this drawback, the main reason for the popularity of ordinary least squares is its computational simplicity, in particular for linear models. The computations are quick and are guaranteed to produce the optimal solution. In contrast, LTS regression aims to minimize the sum of squares of the smallest q percent of residuals, with $50 < q \leq 100$. Among the remaining $100 - q$ percent of residuals, there may be outlying values, but they do not affect the fit result. q can be chosen by the user; the lower limit is 50% (Rousseuw and Leroy, 1987). So why don't we always use LTS with the maximally robust value $q = 50\%$? One drawback is that it effectively discards some of the data, so the precision of the resulting parameter estimates is lower than with ordinary least sum of squares. However, in the context of microarray applications, where there are many thousands of data points, this is rarely a practical concern. A more significant drawback of LTS is that it makes the optimization landscape more rugged. Depending on the data, there are more local minima, and there is a higher risk of getting stuck in these. This can also, sometimes, happen to VSN and therefore you should always carefully check the results of VSN.

What can you do when you do not want to, or cannot, rely on the "most genes not differentially expressed" assumption and the robustness of the VSN estimation algorithm? The recommended strategy in these cases is to identify a subset of features on the array for which this assumption can be assumed to hold, fit the model to these, and then apply it to all features. This subset could, for example, consist of features that target externally spiked in RNA, which has no relation to the biological samples, and is only used for calibration. To demonstrate this, let us assume that we know that 100 features match targets which are equally abundant in both samples.

```
> normctrl = sample(which(!sel), 100)
> fit = vsn2(kidspike[normctrl, ], lts.quantile=1)
> vkidctrl = predict(fit, newdata=kidspike)
> ma = getMA(vkidctrl)
> ma$group = factor("other", levels=c("other",
      "normalization control"))
> ma$group[normctrl] = "normalization control"
```

```
> lp = xyplot(M ~ A , group=group, data=ma,
      pch=".", ylim=c(-3,3), xlim=c(4,16), auto.key=TRUE,
      panel = function(...){
          panel.xyplot(...)
          panel.abline(h=0)})
> print(lp)
```

The result is shown in Figure 5.9.

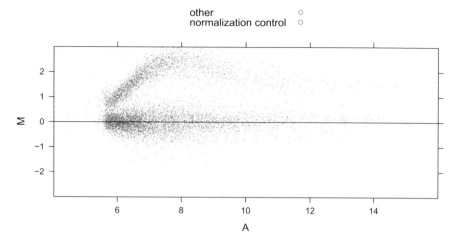

Figure 5.9. The same data as in Figure 5.8, however, now only a set of 100 features ("spike in controls") was used to fit the vsn model, which was then applied to all data.

5.6   Single-color normalization

VSN can also be used for the joint normalization of multiple one-color arrays. For the sake of demonstration, let us pretend that the green color channels of the six arrays of the CC14 dataset form a set of six single-color arrays,

```
> esG  = channel(CC14s, "G")
> exprs(esG) = exprs(esG) - exprs(channel(CC14s, "Gb"))
```

and call justvsn on the *ExpressionSet* object esG.

```
> nesG = justvsn(esG)
```

To assess the result, we can look at the intensity distributions on the arrays before

```
> library("RColorBrewer")
> startedlog = function(x) log2(x+5)
> colors = brewer.pal(6, "Dark2")
> multiecdf(startedlog(exprs(esG)), col=colors,
      main="Before normalization")
```

and after normalization,

```
> multiecdf(exprs(nesG), col=colors,
      main="After normalization")
```

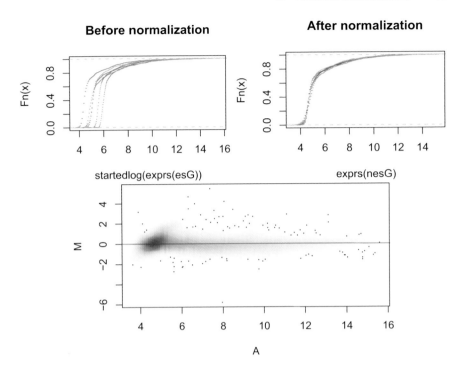

Figure 5.10. The two panels in the upper row show the empirical cumulative distribution functions of the six arrays in esG (before normalization, left) and nesG (after normalization, right). The panel in the bottom row shows the MA-plot between a pair of arrays.

and at the MA-plot between pairs of arrays.

```
> smoothScatter(getMA(nesG[, 1:2]))
> abline(h=0, col="red")
```

The result is shown in Figure 5.10.

5.7 The interpretation of glog-ratios

In Section 5.2, we explored how fold-change estimation might be improved by suitable background-correction and the problem that this poses to the naive $\log_2$-ratio estimator. One possible solution is the generalized logarithm transformation. In this section, we want to study the so-called variance–bias trade-off a bit further. Let us do the following computational experiment. First, we create two vectors g and r with the green and red intensities from the kidney dataset, respectively.

```
> g = exprs(kidney)[,1]
> r = exprs(kidney)[,2]
```

Then we create a small set of 29 artificial "spike-in" data points with A-values from 2 to 16, and M-values of 1, which corresponds to a fold-change of 2.

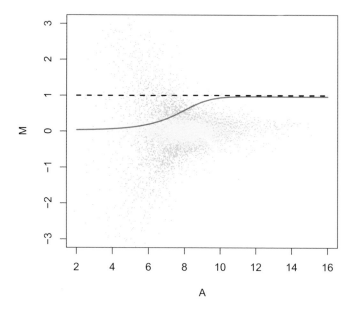

Figure 5.11. The variance–bias trade-off and VSN's property as a shrinkage estimator. Shown is an MA-plot of the kidney data. Dark dots correspond to the naive $\log_2$-ratio, as in Equation (5.1), and light dots to the glog_2-ratio, as in Equation (5.5). The two samples that are compared in these data were taken from immediately adjacent pieces of tissue, so most of the genes are not differentially expressed and have a true log fold-change of 0. Accordingly, both the naive $\log_2$-ratio and the glog_2-ratio are distributed around zero. However, for the naive $\log_2$-ratio, the width of the distribution is bigger for small values of the average intensity A, as can be seen from the "rocket shape" of the distribution. For the glog_2-ratio, the width is approximately constant throughout the range of A. (The visually apparent widening in the intermediate range around $A = 8$ is solely due to the larger density of data points; see Figure 5.7 for a visualization that avoids this artifact.) The lines in this plot are drawn between a set of 29 data points which we have artificially "spiked in" to have a naive $\log_2$-ratio of $\log_2(2) = 1$, at various values of A. This demonstrates the shrinkage effect of VSN: for low intensity data, the glog_2-ratio (solid line), as an estimator of fold-change, shrinks towards zero, but maintains a constant small variance. In contrast, the naive $\log_2$-ratio (dashed line) is unbiased, but its variance increases for low average intensities A.

```
> Aspike = 2^seq(2, 16, by=0.5)
> sel = sample(nrow(kidney), length(Aspike))
> r[sel] = Aspike*sqrt(2)
> g[sel] = Aspike/sqrt(2)
```

Now we can compute A, the naive $\log_2$-ratio M.naive, as well as the glog$_2$-ratio M.vsn.

```
> A = (log2(r) + log2(g))/2
> M.naive =  log2(r) - log2(g)
> fit = vsn2(cbind(g, r))
> M.vsn = exprs(fit)[,2] - exprs(fit)[,1]
```

The plot created by the following lines is shown in Figure 5.11.

```
> plot(A, M.naive, pch=".", ylab="M",
    xlim=c(2,16), ylim=c(-3,3), col="grey")
> points(A, M.vsn, pch=".", col="mistyrose")
> sel = sel[order(A[sel])]
> lines(A[sel], M.vsn[sel], col="red", lwd=2)
> lines(A[sel], M.naive[sel], lwd=2, lty=2)
```

For another analysis of the topic of this section, please refer to the section *VSN, shrinkage, and background correction* in the **vsn** package's vignette *Introduction to robust calibration and variance stabilisation with vsn*.

5.8   Reference normalization

So far, we have been using VSN to normalize a set of microarrays in order to make them comparable among each other. Sometimes, we have an application where we want to add a further array, or set of arrays, without changing the normalization of the existing set. For example, suppose we have used a set of training arrays for setting up a classifier that is able to discriminate different biological states of the samples based on their mRNA profile. Now we get new arrays to which we want to apply the classifier. Clearly, we do not want to rerun the normalization for the whole, new, and bigger dataset, as this would change the training data; neither can we normalize only the new arrays among themselves, without normalizing them "towards" the reference training dataset. What we need is a normalization procedure that normalizes the new arrays with the existing dataset as a reference and without changing the latter. This is possible with VSN.

Let us use the CC14s example data to explore this. First, we call vsn2 on the first five arrays only.

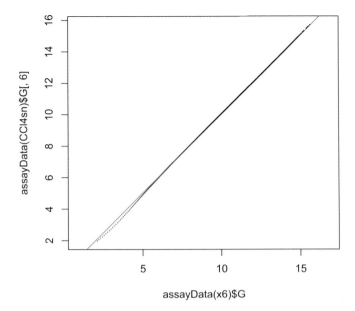

Figure 5.12. Scatterplot of normalized red channel intensities after normalization by reference (x-axis) and joint normalization (y-axis). There is good agreement.

```
> ref = vsn2(CC14s[, 1:5], backgroundsubtract=TRUE)
```

Now we can call justvsn on the sixth array with the output from the previous call as the reference.

```
> x6 = justvsn(CC14s[, 6], reference = ref,
        backgroundsubtract=TRUE)
```

We can compare this to what we got previously, in Section 5.3, for CC14sn.

```
> plot(assayData(x6)$G, assayData(CC14sn)$G[,6], pch=".",
        asp=1)
> abline(a=0, b=1, col="red")
```

The scatterplot in Figure 5.12 shows that the results are approximately the same. For further reading on reference normalization, as well as on some of the other topics of this chapter, please refer to the vignettes *Verifying and assessing the performance with simulated data* and *Likelihood Calculations for vsn* of the **vsn** package, and to the references by Huber et al. (2002, 2003, 2005).

6

Easy Differential Expression

F. Hahne and W. Huber

Abstract

In this short exercise, we explore the most basic approach to the selection of differentially expressed genes between two classes: first, a nonspecific filtering step to remove probes for genes that appear to be not. Second, a probe-by-probe statistical test, and third, multiple testing correction. There are many variations and improvements to the procedure shown here, and you can learn more about these in Chapter 7.

6.1 Example data

For this chapter, we use the `ALL` data, which have been obtained in a microarray study of B- and T-cell leukemia. We want to find genes that are differentially expressed between two distinct types of B-cell leukemia.

```
> library("Biobase")
> library("genefilter")
> library("ALL")
> data("ALL")
```

The data and the following steps with which we construct the subset of interest, `ALL_bcrneg`, are described in more detail in Chapter 1. Briefly, we select samples from B-cell lymphomas harboring the BCR/ABL translocation and from lymphomas with no observed cytogenetic abnormalities (NEG).

```
> bcell = grep("^B", as.character(ALL$BT))
> moltyp = which(as.character(ALL$mol.biol)
        %in% c("NEG", "BCR/ABL"))
```

F. Hahne et al., *Bioconductor Case Studies*, DOI: 10.1007/978-0-387-77240-0_6,
© Springer Science+Business Media, LLC 2008

```
> ALL_bcrneg = ALL[, intersect(bcell, moltyp)]
> ALL_bcrneg$mol.biol = factor(ALL_bcrneg$mol.biol)
```

The last line in the code above is used to drop unused levels of the *factor* variable mol.biol.

6.2 Nonspecific filtering

Between these two groups we should be able to detect substantial differences in gene expression. But first let us explore how nonspecific filtering can improve our analysis. To this end, we calculate the overall variability across arrays of each probe set, regardless of the sample labels. For this, we use the function rowSds, which calculates the standard deviation for each row. A reasonable alternative would be to calculate the interquartile range (IQR), for which we could employ the rowQ function from the **genefilter** package.

```
> library("genefilter")
> sds = rowSds(exprs(ALL_bcrneg))
> sh = shorth(sds)
> sh
[1] 0.242
```

We can plot the histogram of the distribution of sds; see Figure 6.1. The function shorth calculates the midpoint of the *shorth* (the shortest interval containing half of the data), and is in many cases a reasonable estimator of the "peak" of a distribution. Its value 0.242 is drawn as a dashed vertical line in Figure 6.1.

```
> hist(sds, breaks=50, col="mistyrose", xlab="standard deviation")
> abline(v=sh, col="blue", lwd=3, lty=2)
```

There are a large number of probe sets with very low variability. We can safely assume that we will not be able to infer differential expression for their target genes. The target genes of these probe sets may not be expressed in the samples, or the probe sets may lack the sensitivity to detect expression. Hence, let us discard those probe sets whose standard deviation is below the value of sh.

```
> ALLsfilt = ALL_bcrneg[sds>=sh, ]
> dim(exprs(ALLsfilt))
[1] 8812   79
```

Histogram of sds

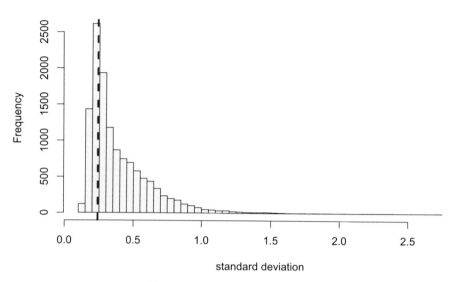

Figure 6.1. Histogram of `sds`.

A related approach would be to discard all probe sets with consistently low expression values. The idea is similar: those probe sets most likely match transcripts whose expression we cannot detect anyway, and hence we need not test them for differential expression.

A more comprehensive approach to nonspecific filtering of probe sets according to various criteria is provided by the function `nsFilter` from the **Category** package, and that function's documentation as well as an application of it in Chapter 1 are further references on this topic.

To summarize, nonspecific filtering uses the biological knowledge that there exists a substantial fraction of probe sets in a microarray experiment that is not informative, either because the target gene is not expressed, or because the probe set lacks sensitivity. Using this knowledge in the analysis will, in general, improve the quality of the gene selection.

6.3 Differential expression

We can now perform probe-by-probe tests for differential expression (Dudoit et al., 2002). The function `rowttests` can deal with *ExpressionSet*s. It uses the *t*-test, row by row, to detect significant differences in the location of the distribution of expression data of two groups of samples defined by a factor variable. In this case, we use the information about BCR/ABL mutation status in the column `mol.biol` of `ALLsfilt`'s sample annotation as a grouping factor.

```
> table(ALLsfilt$mol.biol)
BCR/ABL     NEG
     37      42
> tt = rowttests(ALLsfilt, "mol.biol")
> names(tt)
[1] "statistic" "dm"          "p.value"
```

Take a look at the histogram of the resulting p-values in the left panel of Figure 6.2.

```
> hist(tt$p.value, breaks=50, col="mistyrose", xlab="p-value",
      main="Retained")
```

We see a number of probe sets with very low p-values (which correspond to differentially expressed genes) and a whole range of insignificant p-values. This is more or less what we would expect. The expression of the majority of genes is not significantly shifted by the BCR/ABL mutation. To make sure that the nonspecific filtering did not throw away an undue amount of promising candidates, let us take a look at the p-values for those probe sets that we filtered out before. We can compute t-statistics for them as well and plot the histogram of p-values (right panel of Figure 6.2):

```
> ALLsrest = ALL_bcrneg[sds<sh, ]
> ttrest = rowttests(ALLsrest, "mol.biol")
> hist(ttrest$p.value, breaks=50, col="lightblue",
        xlab="p-value", main="Removed")
```

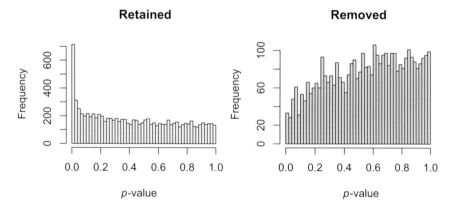

Figure 6.2. Histograms of p-values. The left panel shows those p-values retained after nonspecic lering; the right panel shows those that were removed.

Exercise 6.1
Comment on the plot; do you think that the nonspecific filtering was appropriate?

6.4 Multiple testing correction

We use the *p*-values for ranking genes, and do not advocate interpreting them as true probabilities. Nevertheless, the results of a multiple testing adjustment can be informative for choosing selection cut-offs. Typically, in the setting of a single statistical test we consider the data as providing evidence against a given null hypothesis when it is sufficiently improbable that these data arise by chance if the null hypothesis is true. When repeatedly doing tests, we need to raise the bar for what we consider "sufficiently improbable".

For example, if we do 8812 tests of a null hypothesis that is actually true, using a significance level of 5%, then in 5% ≈ 441 cases we can expect to reject the null hypothesis just by chance. Many approaches have been proposed to address this problem (Pollard et al., 2005); here we just discuss one that appears to be appropriate in many micrarray-related contexts: the false discovery rate (FDR), that is, the expected proportion of false positives among the genes that are called differentially expressed. The procedure of Benjamini and Hochberg is implemented in the **multtest** package and we use the function `mt.raw2adjp` for this purpose. (Note that a more formal treatment would need to take into account the multiple *t*-tests as well as the implicit testing of the nonspecific filtering.)

```
> library("multtest")
> mt = mt.rawp2adjp(tt$p.value, proc="BH")
```

Finally, we can use the results of the *t*-tests to create a gene list containing the ten highest-ranking genes with respect to the adjusted *p*-value,

```
> g = featureNames(ALLsfilt)[mt$index[1:10]]
```

print their gene symbols,

```
> library("hgu95av2.db")
> links(hgu95av2SYMBOL[g])
      probe_id  symbol
1       1635_at    ABL1
2     1636_g_at    ABL1
3       1674_at    YES1
4      32434_at  MARCKS
```

```
5      37015_at ALDH1A1
6      37027_at   AHNAK
7      39730_at    ABL1
8    39837_s_at  ZNF467
9      40202_at    KLF9
10     40504_at    PON2
```

and plot the data of the first one together with symbols indicating the value
of the `mol.biol` variable:

```
> mb  = ALLsfilt$mol.biol
> y   = exprs(ALLsfilt)[g[1],]
> ord = order(mb)
> plot(y[ord], pch=c(1,16)[mb[ord]],
      col=c("black", "red")[mb[ord]],
      main=g[1], ylab=expression(log[2]~intensity),
      xlab="samples")
```

The result is shown in Figure 6.3.

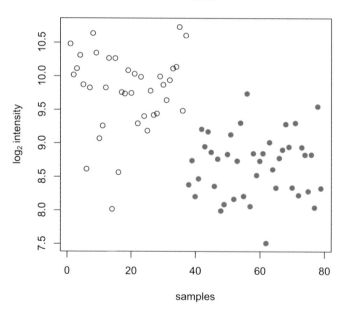

Figure 6.3. The ALLsfilt data for the top differentially expressed probe set across
the 79 samples. The value of the `mol.biol` variable is indicated by the plot
symbols.

7

Differential Expression

W. Huber, D. Scholtens, F. Hahne, and A. von Heydebreck

Abstract

In this chapter we will cover some of the basic principles of finding differentially expressed genes. We cover nonspecific filtering, multiple testing, the moderated test statistics provided by the **limma** package, and gene selection by ROC curves.

7.1 Motivation

There are many different ways to detect differentially expressed genes. Rather than prescribe a standard way of doing this, in this chapter we explore a variety of these. The goal is to give you an overview of the existing ideas so you can make an appropriate choice in your own analyses.

7.1.1 The gene-by-gene approach

In current practice, differential expression analysis is generally done using a gene-by-gene approach. This ignores the dependencies between genes, for example their interrelationships in regulatory modules. Clearly, this is not satisfactory, and it will change as we learn more. For the purpose of this chapter, we cover the gene-by-gene approach.

7.1.2 Nonspecific filtering

Most microarrays contain probes for many more genes than will be differentially expressed. Indeed, one of the basic assumptions of normalization is that most genes are not differentially expressed. To alleviate the loss of power from the formidable multiplicity of gene-by-gene hypothesis testing, we advise that some form of nonspecific prefiltering should be carried out.

F. Hahne et al., *Bioconductor Case Studies*, DOI: 10.1007/978-0-387-77240-0_7,
© Springer Science+Business Media, LLC 2008

By *nonspecific* we mean that it is done without reference to the parameters or conditions of the tested RNA samples. Its aim is to remove from consideration that set of probes whose genes are not differentially expressed under any comparison. We have found it most useful to select genes on the basis of variability (von Heydebreck et al., 2004). Only the genes that show any variation across samples can potentially be differentially expressed among our groups of interest.

7.1.3   Fold-change versus t-test

The simplest approach is to select genes using a fold-change criterion. This may be the only possibility in cases where few replicates are available. An analysis solely based on fold-change, however, precludes the assessment of significance of observed differences in the presence of biological and experimental variation, which may differ from gene to gene. This is the main reason for using statistical tests to assess differential expression.

In general, one might look at all sorts of differences between the distributions of a gene's expression levels under different conditions. Most often, the location (e.g., mean or median) parameter is considered. This leads to the *t*-test and its variations. There are also good reasons to consider other properties of the distributions, such as the partial area under the Receiver Operating Characteristic (ROC) curve of a threshold classifier (Pepe et al., 2003).

One may distinguish between parametric tests, such as the *t*-test, and nonparametric tests, such as the Mann–Whitney test or permutation tests. Parametric tests usually have a higher power if the underlying model assumptions, such as Normality in the case of the *t*-test, are at least approximately fulfilled. Nonparametric tests have the advantage of making less stringent assumptions on the data-generating distribution. In many microarray studies, however, a small sample size leads to insufficient power for nonparametric tests. A pragmatic approach in these situations is to employ parametric tests, but to use the resulting *p*-values cautiously to rank genes by their evidence for differential expression.

7.2   Nonspecific filtering

Let us load the dataset which we work on. In Chapter 1 you can find a comprehensive description of the acute lymphoblastic leukemia data that we use here.

```
> library("ALL")
> data("ALL")
```

First, we construct a list of samples from tumors of B-cells.

```
> bcell = grep("^B", as.character(ALL$BT))
```

The BCR/ABL translocation – formally, t(9;22)(q34;q11), which is often called the Philadelphia chromosome, producing a fusion gene consisting of the BCR and the ABL1 genes – is relatively prominent in acute lymphocytic leukemias and of therapeutic relevance. Here, we focus on the subset of ALL samples that harbor this translocation and contrast it with the group of samples for which none of the common cytogenetic aberrations (group NEG) was detected.

```
> moltyp = which(as.character(ALL$mol.biol)
      %in% c("NEG", "BCR/ABL"))
```

Let us now construct a new data object ALL_bcrneg that contains only those samples that fulfill these two conditions.

```
> ALL_bcrneg = ALL[, intersect(bcell, moltyp)]
> ALL_bcrneg$mol.biol = factor(ALL_bcrneg$mol.biol)
```

The second line of the above code chunk cleans up the *factor* variable ALL_bcrneg$mol.biol by removing the empty levels.

Now, if we are going to filter on the basis of variability, we might first want to make sure that the variability is not dominated by its dependence on the mean expression level. If it were, then selecting on the basis of variability would be confounded with selection on the basis of absolute level. There are good reasons, in essence due to the existence of probe-sequence specific background and gain factor effects, not to use the absolute level for gene selection. To check for an association we plot rowwise means versus rowwise standard deviations and plot these together with a smoothed estimate of their regression.

```
> library("vsn")
> meanSdPlot(ALL_bcrneg)
```

The result is shown in Figure 7.1.

Exercise 7.1
  a. *Comment on the plot; do you think that the relationship between mean and standard deviation is sufficiently weak?*

  b. *Have a look at the manual page of meanSdPlot. What is the use of the ranks parameter?*

Presuming that we decide that the relationship is not very strong, we proceed. Our next step is to set aside those probe sets with low variability.

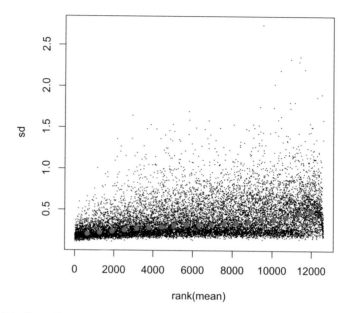

Figure 7.1. Rowwise means versus rowwise standard deviations of the ALL data.

In the code below, we set aside the 80% lowest variability probe sets. We choose such a high fraction because we want to limit the length of the subsequent computations. The best choice for this fraction depends on the array design and the biological samples, but in practice it will usually be much lower.

```
> sds = esApply(ALL, 1, sd)
> sel = (sds > quantile(sds, 0.8))
> ALLset1 = ALL_bcrneg[sel,]
```

A potential drawback of this approach is found in situations where we are interested in an experimental factor in which one group of samples has few members. In this case, a gene which is differentially expressed between that group and the other(s) may not have a large overall standard deviation. How would you address this situation?

At this point you may want to try to look at some heatmaps of the data to see if there are any obvious patterns. Consult the manual page of the function by typing: ?heatmap.

7.3 Differential expression

In Bioconductor, the **genefilter** package allows you to easily select genes using a variety of filters. Additionally, for some tests and comparisons we have developed fast versions. These include rowttests, which perform a *t*-test for every row in a gene expression matrix; `rowFtests`, which does *F*-tests; and `rowQ`, which calculates a quantile for each row.

First, and perhaps easiest is to use a *t*-test (Dudoit et al., 2002).

```
> library("genefilter")
> tt = rowttests(ALLset1, "mol.biol")

> names(tt)
[1] "statistic" "dm"         "p.value"
```

Consult the manual page for `rowttests` for the meaning of the four different elements of the return value `tt`. Many practitioners have learned that small *p*-values do not always correspond to genes for which there have been large changes. Let us look at the so-called volcano plot.

```
> plot(tt$dm, -log10(tt$p.value), pch=".",
      xlab = expression(mean~log[2]~fold~change),
      ylab = expression(-log[10](p)))
```

The result is shown in Figure 7.2.

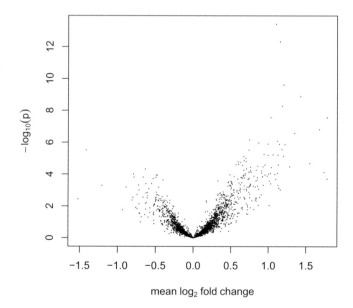

Figure 7.2. Volcano plot. Plotted along the *x*-axis is a measure of effect size (here, the mean fold-change), along the *y*-axis a measure of significance (here, the negative logarithm of the *p*-value). Some of the data points have a large average fold-change, but low significance, and high significance does not always imply large effect size.

Exercise 7.2
Determine how many probe sets correspond to differentially expressed genes, using the t-test results.

7.4   Multiple testing

One of the subject areas that has received a great deal of attention is that of multiple testing. We provide a brief introduction to the functionality in the **multtest** package (Pollard et al., 2005).

Many of the algorithms in the **multtest** package depend on random permutations of the samples. The number of permutations is controlled by the parameter B. In the following, we call mt.maxT to perform a permutation test, using the Welch statistic.

```
> library("multtest")
> cl = as.numeric(ALLset1$mol.biol=="BCR/ABL")
> resT = mt.maxT(exprs(ALLset1), classlabel=cl, B=1000)
> ord = order(resT$index)    ## the original gene order
> rawp = resT$rawp[ord]      ## permutation p-values
```

Figure 7.3 shows the histogram of unadjusted permutation p-values as given by the vector rawp. The high proportion of small p-values suggests that indeed a substantial fraction of the genes is differentially expressed between the two groups.

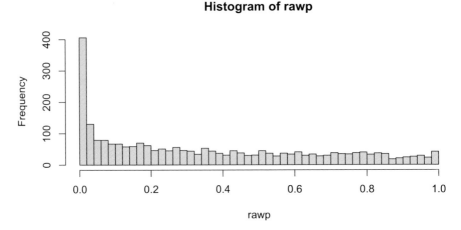

Figure 7.3. Histogram of the unadjusted p-values.

```
> hist(rawp, breaks=50, col="#B2DF8A")
```

In order to control the familywise error rate (FWER), that is, the probability of at least one false positive in the set of significant genes, mt.maxT used the permutation-based maxT procedure of Westfall and Young (1993). We obtain 34 genes with an adjusted p-value below 0.05:

```
> sum(resT$adjp<0.05)
[1] 34
```

A comparison of this number to the height of the leftmost bar in the histogram suggests that we are missing a large number of differentially expressed genes. The FWER is a very stringent criterion, and in some microarray studies, few or no genes may be significant in this sense, even if many more are truly differentially expressed. A more sensitive criterion is provided by the false discovery rate (FDR), that is, the expected proportion of false positives among the genes that are called significant. We can use the procedure of Benjamini and Hochberg (1995) as implemented in **multtest** to control the FDR:

```
> res = mt.rawp2adjp(rawp, proc = "BH")
> sum(res$adjp[,"BH"]<0.05)
[1] 209
```

7.5 Moderated test statistics and the **limma** package

We use the p-values for ranking genes, and do not advocate interpreting them as true probabilities. Nevertheless, the results of a multiple testing adjustment can be informative for choosing selection cut-offs. Note that a more formal treatment would need to take into account the multiple t-tests as well as the implicit testing of the nonspecific filtering.

A t-test analysis can also be conducted with functions of the **limma** package (Smyth, 2004). First, we have to define the design matrix. One possibility is to use an intercept term that represents the mean $\log_2$ intensity of a gene across all samples (the first column in the below matrix, consisting of 1s), and to encode the difference between the two classes in the second column.

```
> library("limma")
> design  = cbind(mean = 1, diff = cl)
```

Next a linear model is fitted for every gene by the function lmFit, and an empirical Bayes moderation of the standard errors can be performed with the function eBayes (Smyth, 2004). This employs information from all genes to arrive at more stable estimates of each individual gene's variance.

```
> fit = lmFit(exprs(ALLset1), design)
> fit = eBayes(fit)
```

We can list the ten most differentially expressed genes using the function topTable. The three probe sets with the lowest p-value all map to the ABL1 gene which is part of the fusion gene product caused by the t(9;22)(q34;q11) translocation and which is known to be over-expressed and acting as a strong oncogene in acute lymphoblastic leukemia.

```
> library("hgu95av2.db")
> ALLset1Syms = unlist(mget(featureNames(ALLset1),
        env = hgu95av2SYMBOL))
> topTable(fit, coef = "diff", adjust.method = "fdr",
        sort.by = "p", genelist = ALLset1Syms)
          ID logFC AveExpr     t   P.Value adj.P.Val      B
156     ABL1 1.100   9.196 9.034 4.879e-14 1.232e-10 21.293
1915    ABL1 1.153   9.000 8.588 3.877e-13 4.895e-10 19.341
155     ABL1 1.203   7.897 7.339 1.229e-10 1.034e-07 13.906
163     YES1 1.427   5.002 7.050 4.554e-10 2.875e-07 12.668
2066    PON2 1.181   4.244 6.665 2.571e-09 1.298e-06 11.032
2014    KLF9 1.779   8.621 6.392 8.623e-09 3.629e-06  9.889
1262 ALDH1A1 1.033   4.331 6.242 1.662e-08 5.997e-06  9.269
437   MARCKS 1.679   4.466 5.972 5.376e-08 1.697e-05  8.162
1269   AHNAK 1.349   8.444 5.805 1.097e-07 3.078e-05  7.489
1366   ANXA1 1.118   5.087 5.483 4.265e-07 1.077e-04  6.211
```

When you compare the resulting p-value with those from the parametric t-test, you will see that they are almost identical:

```
> plot(-log10(tt$p.value), -log10(fit$p.value[, "diff"]),
        xlab = "-log10(p) from two-sample t-test",
        ylab = "-log10(p) from moderated t-test (limma)",
        pch=".")
> abline(c(0, 1), col = "red")
```

The result is shown in Figure 7.4. Because of the large number of samples, the empirical Bayes moderation is not so relevant here: in these dataset the gene-specific variance can be well estimated from the data of each gene.

7.5.1 Small sample sizes

However, the empirical Bayes moderation may be quite useful in cases with fewer replicates. Let us draw a subsample with three arrays from each group from our data:

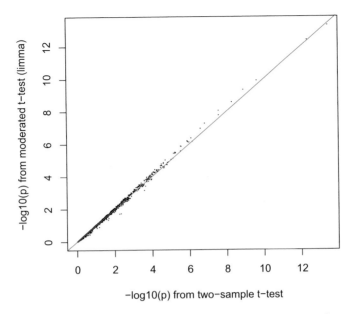

Figure 7.4. Comparison of *p*-values from the unmoderated and moderated *t*-test in a situation with a large number of samples in both groups.

```
> subs = c(35, 65, 75, 1, 69, 71)
> ALLset2 =  ALL_bcrneg[, subs]
> table(ALLset2$mol.biol)
BCR/ABL     NEG
      3       3
```

We repeat the testing procedure in the same way as before,

```
> tt2  = rowttests(ALLset2, "mol.biol")
> fit2 = eBayes(lmFit(exprs(ALLset2), design=design[subs, ]))
```

and plot the results in Figure 7.5.

```
> plot(-log10(tt2$p.value), -log10(fit2$p.value[, "diff"]),
      xlab = "-log10(p) from two-sample t-test",
      ylab = "-log10(p) from moderated t-test (limma)",
      pch=".")
> abline(c(0, 1), col = "red")
```

Let us have a look at a gene that has a small *p*-value in the normal *t*-test but a large one in the moderated test.

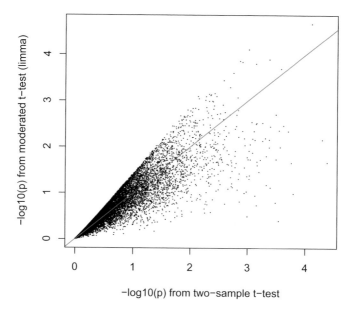

Figure 7.5. Comparison of p-values from the unmoderated and moderated t-test in a situation with a small number of samples in both groups.

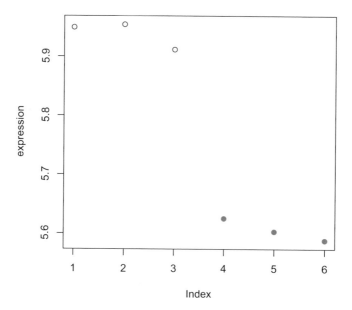

Figure 7.6. Expression values of a probe set that has a highly significant p-value in the unmoderated t-test, but is unremarkable in the moderated test. Note the y-axis scaling: although the two groups look well separated, the absolute difference is small. Most likely, this is a chance artifact. You can verify this by looking at the expression values of this probe set in the other samples that were not used here.

```
> g = which(tt2$p.value < 1e-4 &
      fit2$p.value[, "diff"] > 0.02)
```

We plot its expression values and use different symbols and colors to encode the sample classes.

```
> sel = (ALLset2$mol.bio == "BCR/ABL")+1
> col = c("black", "red")[sel]
> pch = c(1,16)[sel]
> plot(exprs(ALLset2)[g,], pch=pch, col=col,
      ylab="expression")
```

The plot is shown in Figure 7.6.

7.6   Gene selection by Receiver Operator Characteristic (ROC)

In this section we consider a method for finding differentially expressed genes that, instead of the hypothesis testing approaches we have explored so far, uses a classification-based approach. The approach was presented by Pepe et al. (2003).

The aim is to find genes that might serve as potential markers, that is, genes whose expression level, individually, is able to discriminate between two groups. Let us call one group Control and the other one Disease. If we denote by x the observed expression level of a given gene, we consider the simple classification rule that all samples with $x \geq \theta$, for some choice of θ, are predicted to be in the Disease class and all samples with $x < \theta$ in the Control class. For different choices of θ, the performance of the classifier can be measured by its specificity, that is, the probability that a true Control sample is classified as a Control, and by its sensitivity, the probability that a true Disease sample is classified as Disease. The plot of sensitivity versus $p = 1 - $ specificity is called the Receiver Operator Characteristic (ROC) curve. An example is shown in Figure 7.7.

We would like to identify genes that have the best ability for detecting whether a sample has the BCR/ABL translocation. This can be expressed by the area under the ROC curve (AUC), or more generally, the pAUC. The pAUC criterion, for a small value of p such as $p = 0.2$, is often more relevant than the AUC because for a practical diagnostic marker we will require high specificity, say, better than 80%, before even considering its sensitivity.

Similar to rowttests, the package **genefilter** contains a function for row-wise computation of the pAUC statistics. As arguments, the function takes

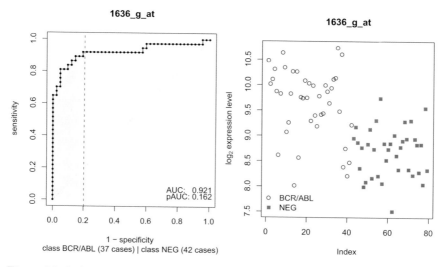

Figure 7.7. Left: ROC curve of a probe set that separates well between BCR/ABL positive and negative tumors. The (total) area under the curve (AUC) is the area of the region surrounded by the x-axis, the line $x = 1$, and the plotted curve. The partial area under the curve (pAUC) for $p = 0.2$ is the shaded area surrounded by the x-axis, the line $x = 0.2$, and the curve. Right: The probe set's data, sorted and colored by the value of the molecular subtype.

an *ExpressionSet* object x, a character giving the name of the factor covariate and the numeric scalar p. In addition, the function has an argument flip. If it is set to TRUE (the default), then for each gene both classification rules $x < \theta$ and $x > \theta$ are tested, and the (partial) area under the curve of the better one of the two is returned. For our data this makes sense, because the Disease classification is somewhat arbitrary and not necessarily linked to higher expression values. Instead, we expect to find both over- and underexpressed genes for each class. You can set flip to FALSE if you only want to screen for genes which discriminate Disease from Control with the $x > \theta$ rule. For more advanced classification rules you might want to consider the functionality provided by the **ROC** package.

```
> rocs = rowpAUCs(ALLset1, "mol.biol", p=0.2)
```

Next we select the probe set with the maximal value of our pAUC statistic, and plot the corresponding ROC curve.

```
> j = which.max(area(rocs))
> plot(rocs[j], main = featureNames(ALLset1)[j])
```

The result is shown in the left panel of Figure 7.7.

Exercise 7.3
Plot the expression values of `1636_g_at`, *as in the right panel of Figure 7.7.*

Exercise 7.4
How would you expect the ROC curve to look for a gene that does not show any differential expression between the groups? How would an "ideal" ROC curve look? Why?

7.7   When power increases

In Sections 7.3 and 7.6 we have shown how to find differentially expressed genes using two different criteria: location changes in the distribution of a gene's expression and the receiver operating characteristic curve of a simple threshold classifier. Let us now compare these methods and focus on the influence of sample size.

   We want to see how the sample size affects the number of genes that are found to be differentially expressed in the two methods. For this, we repeatedly take random sample subsets of our data, of varying size, and do the differential expression computations on each of these subsets, each time resulting in a certain number of differentially expressed genes.

   Let us first build wrapper functions around `rowttests`

```
> nrsel.ttest = function(x, pthresh=0.05) {
      pval = rowttests(x, "mol.biol")$p.value
      return(sum(pval < pthresh))
  }
```

and `rowpAUCs`.

```
> nrsel.pAUC = function(x, pAUCthresh=2.5e-2) {
      pAUC = area(rowpAUCs(x, fac="mol.biol", p=0.1))
      return(sum(pAUC > pAUCthresh))
  }
```

   Note that the choices of thresholds are, as always, somewhat arbitrary, and that the one for the *t*-test, `pthresh`, is not directly comparable to the one for `rowpAUCs`, `pAUCthresh`.

   What we need now is a function that does resampling for various dataset sizes and plots the output. The code for this is straightforward, but somewhat tedious, so for simplicity we put a function that does this for you into the package **BiocCaseStudies**; it is called `resample`. If you want you can take a look at its code by typing `resample`.

   Because the following computation can take a lot of time, for the purpose of demonstration we construct a subset of `ALLset1` with only 1000 probes.

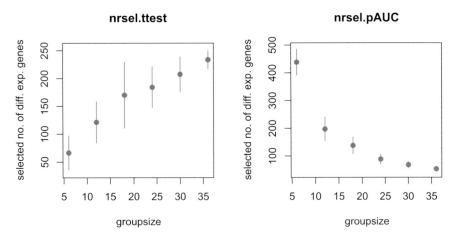

Figure 7.8. Comparing the number of differentially expressed genes for different sample sizes. Left: *t*-test criterion. Right: pAUC criterion.

(If you are working on a powerful computer, you could call the following code with `ALLset1` instead of x.)

```
> library(BiocCaseStudies)
> x = ALLset1[sample(nrow(ALLset1), 1000), ]
```

Run the resampling function, once with `nrsel.ttest` and once with `nrsel.pAUC`. You need to pass along x as well. Take a look at the output in Figure 7.8.

```
> resample(x, "nrsel.ttest")
> resample(x, "nrsel.pAUC")
```

Exercise 7.5
Which of the two criteria seems more reasonable?

Topics for further study:

- In this vignette you have seen a number of different ways to determine differentially expressed genes. What other ways are there to select interesting genes?

- How do the results from the different methods you have seen here overlap? What is the reason in those cases where they produce different results?

- How could you formulate criteria to decide which selection method is better (or more appropriate for your problem) than the others?

- How do the different methods deal with outliers and with missing data?

8

Annotation and Metadata

W. Huber and F. Hahne

Abstract

In this chapter we demonstrate the use of Bioconductor metadata resources. After having obtained a list of reporters from a microarray experiment and mapping them to their target genes, one will want to use the annotation of the genes and gene products to better interpret the experimental results. Often, it is beneficial to use gene annotation in the course of the primary analysis, in order to narrow down the set of data to be considered and ameliorate multiple testing problems, or in order to explore specific biological hypotheses.

8.1 Our data

We make use of a subset of the ALL data. We refer you to Chapter 1 for a comprehensive description of these data and of the following code, in which we load the data and select those samples that were obtained from tumors harboring either the BCR/ABL or the ALL1/AF4 translocation.

```
> library("ALL")
> data("ALL")
> types = c("ALL1/AF4", "BCR/ABL")
> bcell = grep("^B", as.character(ALL$BT))
> ALL_af4bcr = ALL[, intersect(bcell,
      which(ALL$mol.biol %in% types))]
> ALL_af4bcr$mol.biol = factor(ALL_af4bcr$mol.biol)
```

We want to apply a nonspecific filtering step in order to remove probe sets that are likely to be noninformative. We use the function nsFilter from the **genefilter** package for that purpose. The default measure used by nsFilter for the variance filtering step is the IQR. This is a reasonable choice as long as the sizes of the sample groups are approximately similar. This is not the

F. Hahne et al., *Bioconductor Case Studies*, DOI: 10.1007/978-0-387-77240-0_8,

case for our BCR/ABL ALL1/AF4 subset, where the ALL1/AF4 positive group is much smaller:

```
> table(ALL_af4bcr$mol.biol)
ALL1/AF4  BCR/ABL
      10       37
```

For calculation of the IQR, 50% of the most extreme values are discarded as outliers, thus the measure of variance will be dominated mainly by the much larger BCR/ABL positive group. We could address this problem by using a non-robust measure of variance like the standard deviation, however this would make the filtering more susceptible to outliers. Instead, we will look at the range between more extreme quantiles, here.

```
> qrange <- function(x)
      diff(quantile(x, c(0.1, 0.9)))
> library("genefilter")
> filt_af4bcr = nsFilter(ALL_af4bcr, require.entrez=TRUE,
      require.GOBP=TRUE, var.func=qrange, var.cutoff=0.5)
> ALLfilt_af4bcr = filt_af4bcr$eset
```

Now, let us load the packages with the necessary tools and annotation data.

```
> library("Biobase")
> library("annotate")
> library("hgu95av2.db")
```

Our first step is to use the function rowttests to carry out a two-group comparison and to select the top 100 genes.

```
> rt = rowttests(ALLfilt_af4bcr, "mol.biol")
> names(rt)
[1] "statistic" "dm"        "p.value"
```

Exercise 8.1
Plot histograms of the t-statistic and of the p-values, such as in Figure 8.1.

Exercise 8.2
Create an ExpressionSet ALLsub *with the 400 probe sets with smallest p-values.*

There are many variations and possible improvements to this probe set selection procedure, but that is not the goal of this chapter. Here, the goal is to get a reasonable list of probe sets and to subsequently use that to demonstrate the use of metadata.

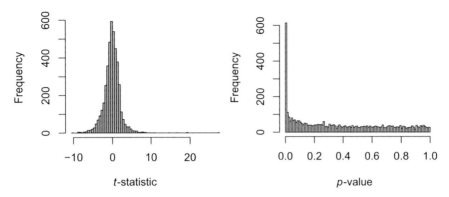

Figure 8.1. Histograms of *t*-statistics and *p*-values.

Exercise 8.3
How many probe sets in `ALL` and how many probe sets in `ALLsub` map to the same EntrezGene ID?

Exercise 8.4
Plot the expression profile of the CD44 gene, as in the left panel of Figure 8.2.

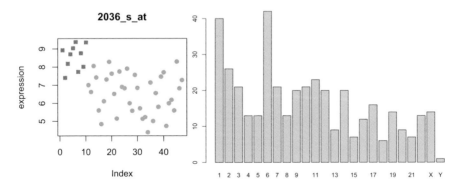

Figure 8.2. The left panel shows the expression profile of a probe set from the `ALLsub` data. Square plot symbols correspond to the samples with the ALL1/AF4 molecular subtype ands circle symbols to BCR/ABL. The right panel is a barplot of frequencies with which the target genes of `ALLsub` map to the different human chromosomes.

Exercise 8.5
Produce a barplot, as in the right panel of Figure 8.2, that indicates for each chromosome the number of genes probed by ALLsub *that are on that chromosome.*

In the code chunk below, we show how to use the **annaffy** package to produce a HTML table for a list of genes, in this case for the 400 genes from the ALLsub.

```
> library("annaffy")
> anncols = aaf.handler(chip="hgu95av2.db")[c(1:3, 8:9, 11:13)]
> anntable = aafTableAnn(featureNames(ALLsub),
      "hgu95av2.db", anncols)
> saveHTML(anntable, "ALLsub.html",
      title="The Features in ALLsub")
> localURL = file.path("file:/", getwd(), "ALLsub.html")
```

We can point our HTML browser to this file.

```
> browseURL(localURL)
```

8.2   Multiple probe sets per gene

The annotation package **hgu95av2.db** provides information about the genes represented on the array, including their EntrezGene identifiers[1], Unigene cluster identifiers, gene names, chromosomal location, Gene Ontology annotation, and pathway associations (Wheeler et al., 2007; Mulder et al., 2007). Although the term *gene* has many aspects and can mean different things to different people, we operationalize it by identifying it with entries in the Entrez database (Maglott et al., 2007). One problem that arises is that some genes are represented by multiple probe sets on the chip.

There are no easy answers to the questions that stem from this, not least because biology is more complex than the one gene–one transcript model that underlies the design of arrays such as the one considered here (ENCODE Project Consortium et al., 2007). In some cases, you may wonder how to handle the fact that one probe set for a gene shows a certain pattern, and another one shows a different pattern. In other cases, you might have several probe sets for one gene, but only one for another, yet that imbalance should not affect your inference.

---

[1] http://www.ncbi.nlm.nih.gov/EntrezGene.

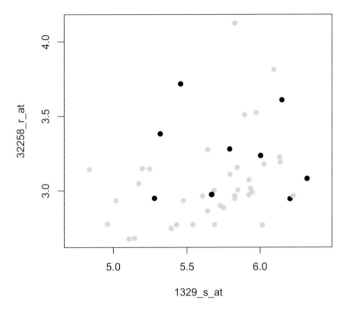

Figure 8.3. Scatterplot of the data from two probe sets that both map to EntrezGene 7013 (TERF1). Dark points correspond to ALL1/AF4 samples.

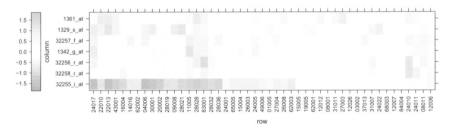

Figure 8.4. Heatmap of the data from seven probes mapping to EntrezGene 7013. Note that the data for each probe set are median centered.

Exercise 8.6
Select some pairs of probe sets that are mapped to the same gene and plot their expression values against each other. You can use Figures 8.3 and 8.4 as examples.

8.3   Categories and overrepresentation

A bit later in this chapter we consider using Gene Ontology annotation data to try to find biological themes in the list of selected genes (The Gene Ontology Consortium, 2000). In this section, we first address a more general question. Suppose that you can divide your genes into k groups. For example, k might be 24 and represent chromosomes in humans. The selection of differentially expressed genes defines another grouping: those that have low p-values, and those that do not. These two categorizations can be used to define a two-way contingency table. In our example it has 24 rows and two columns, and each table element contains the number of genes that falls into the corresponding categories.

The question is now whether the two categorizations (membership on a particular chromosome, and being or not being differentially expressed) are associated. There are different tests for association in two-way tables, Pearson's χ^2 test, provided by the R function chisq.test, and Fisher's exact test, provided by the function fisher.test, among them.

In general, we need to pay attention when doing such analyses to the fact that there can be multiple probe sets per gene. Although annotations in the annotation packages are given per probe set, that information is redundant when the probe sets map to the same EntrezGene ID. Conclusions drawn could be misleading if, for example, three redundant pieces of information were treated as three independent pieces of information. For the current set of example data, ALLsub, this is not a problem, due to the prefiltering with nsFilter that we did above.

Here we have adapted a strategy that considers EntrezGene IDs as the primary keys through which different types of annotation are mapped and among which we search for overrepresentation. Depending on the question of interest, a different approach may be needed, for example, when considering annotation that is more closely related to the proteins.

Exercise 8.7
Create a data.frame chr *with two columns* gene_id *and* chromosome *which for each EntrezGene ID contains the chromosome to which it is mapped.*

Exercise 8.8
Create a contingency table for the association of EntrezGene IDs with their chromosome mapping and with being differentially expressed in the ALLfilt_af4bcr *data (remember the vector* EGsub *that you have created earlier). Use the functions* fisher.test *and* chisq.test *to test for association. (You may need to consult the man pages regarding its parameter* simulate.p.value *to make* fisher.test *work for these data.)*

Once we have established that there exists an association between chromosomal location and being differentially expressed between BCR/ABL and ALL1/AF4, we can try to pin down this association more specifically, for example, by considering the residuals.

A sometimes used but generally inappropriate approach is to separately test the contingency tables for each chromosome. This is usually done using a Hypergeometric sampling model (genes are either on a specific chromosome or not, and they are either differentially expressed or not). Note that this is exactly the same as using Fisher's test for the corresponding two-way table, separately chromosome by chromosome.

Optional exercise: Use the Hypergeometric distribution to consider each chromosome separately. How do these results compare with those found above? Report the per-chromosome summary statistics. If any *p*-values are significant, are there more or fewer genes than what you would expect by chance?

8.3.1   Chromosomal location

In other settings we might be interested in where the genes are located, what other genes are nearby, and perhaps in grouping genes by their location before testing for overrepresentation.

Bioconductor annotation packages contain information on chromosomal location in annotation objects with the CHRLOC suffix. The start location is given as an integer number whose absolute value is the distance of the transcription start site from the 5′ end of the chromosome. The values for genes coded on the sense strand have a positive sign; those for genes on the antisense strand have a negative sign.

Exercise 8.9
How many probe sets in ALLsub are on the sense strand?

Another useful concept related to chromosomal location comes from chromosome bands. This information can be obtained from the MAP annotation objects (e.g., hgu95av2MAP). In some cases the exact location is not known, and a range is given. An example of using these for GSEA is given in Chapter 13.

8.4   Working with GO

The Gene Ontology (GO) is a structured vocabulary of terms describing gene products according to molecular function, biological process, and cellular component (The Gene Ontology Consortium, 2000). The *molecular function* of a gene product describes what it can do at the biochemical level but without reference to where or when this activity might occur. The *biological process* of a gene product describes a biological objective to which the gene product contributes. The *cellular component* ontology describes locations, at the levels of subcellular structures and macromolecular complexes. Examples of cellular components include nuclear inner membrane and the ubiquitin ligase complex.

GO terms can be linked by two relationships: *is a* and *part of.* For example, a *nuclear chromosome* is a *chromosome*, and a *nucleus* is part of a *cell.* The ontologies are structured as directed acyclic graphs (DAG). A parent of a term is a more general GO term that precedes it in the DAG, that is, which is linked to it by a chain of *is a* or *part of* relationships. For example, the biological process term *hexose biosynthesis* has two parents, *hexose metabolism* and *monosaccharide biosynthesis.* For precision and conciseness, all indexing of GO terms employs seven-digit tags with prefix GO:, for example, GO:0008094.

There are three basic tasks. One is navigating the hierarchy, determining parents and children of terms and deriving subgraphs of interest from the overall graph. A second is resolving the mapping from a GO tag to its natural language description, and the third is the mapping between GO terms and genes or gene products. There are software tools in the packages described here to allow users to perform these tasks.

The induced GO graph for a collection of genes is the graph that results from taking the union of all GO terms annotated to the genes and also all their parent terms.

Finding parents and children of different terms is handled by using the PARENT and CHILDREN mappings. To find the children of "GO:0008094" we use:

```
> library("GO.db")
> as.list(GOMFCHILDREN["GO:0008094"])
$`GO:0008094`
         isa             isa             isa             isa
 "GO:0004003"    "GO:0015616"    "GO:0033170"    "GO:0043142"
```

We use the term *offspring* to refer to all descendants (children, grandchildren, and so on) of a node. Similarly we use the term *ancestor* to refer to the parents, grandparents, and so on, of a node.

```
> as.list(GOMFOFFSPRING["GO:0008094"])
$`GO:0008094`
[1] "GO:0003689" "GO:0004003" "GO:0015616" "GO:0017116"
[5] "GO:0033170" "GO:0033171" "GO:0043140" "GO:0043141"
[9] "GO:0043142"
```

8.4.1 Functional analyses

The packages **annotate** and **GOstats** provide much of the necessary functionality for working with GO. Other packages you might want to consider for statistical analyses are **topGO** and **goTools**.

The function hyperGTest will compute the Hypergeometric *p*-values for overrepresentation of genes at all GO terms in the induced GO graph. The basic idea is simple. Consider an experimental method that identifies a set of genes as, say, either differentially expressed or not. This set of genes is called the *gene universe*. Then, for a specific GO term, the universe is partitioned into those genes that are annotated at the term and those that are not. You might think of this as an urn, with all genes in the universe represented as balls in the urn. The genes annotated at the GO term are black; the others are white. We can now ask whether the frequency of black balls among the differentially expressed genes is surprisingly high, or whether it is just consistent with the overall frequency of black balls in the universe. If the black balls are overrepresented, one may conclude that there appears to be an association between the GO term and the list of interesting genes.

This test is performed many times, once for each GO category. Because of the way that genes are annotated at GO terms, there are concerns with how one might address the multiple testing issues that arise. Both the **topGO** and **GOstats** packages implement a form of conditional testing that is designed to address some of these concerns. More details on the conditional testing paradigm can be found in Chapter 14 and Falcon and Gentleman (2007). Gene set enrichment analysis (GSEA) is another approach and it is considered in some detail in Chapter 13.

```
> library("GOstats")
```

To perform the test we first define the universe of genes and create a parameter object to set the analysis options.

```
> affyUniverse = featureNames(ALLfilt_af4bcr)
> uniId = hgu95av2ENTREZID[affyUniverse]
> entrezUniverse = unique(as.character(uniId))
> params = new("GOHyperGParams",
      geneIds=EGsub, universeGeneIds=entrezUniverse,
      annotation="hgu95av2", ontology="BP",
      pvalueCutoff=0.001, conditional=FALSE,
      testDirection="over")
```

```
> mfhyper = hyperGTest(params)
```

Coming up with an appropriate approach to adjusting for the multiple testing is not straightforward. The tests are not independent, the p-values are related to the size of each category (i.e., the number of genes annotated to it), and the sampling distribution is not clear. Despite this, many people do use some sort of multiple testing adjustments. We often prefer a pragmatic approach that considers the unadjusted p-values as well as the size of the estimated effects, for example, the odds-ratios. We can plot the histogram of unadjusted p-values; see Figure 8.5.

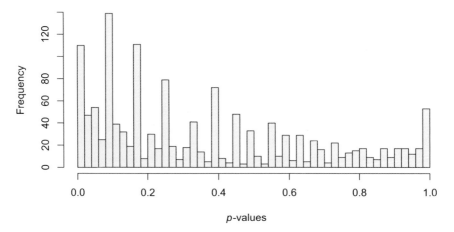

Figure 8.5. Histogram of p-values for overrepresentation of genes at all GO terms in the induced GO graph.

```
> hist(pvalues(mfhyper), breaks=50, col="mistyrose")
```

We see a peak on the left side, close to zero, which indicates that there seem to be some categories that are enriched with genes beyond what is expected by chance.

Exercise 8.10
Look at the GO categories that appear to be significantly overrepresented. You could use a p-value cutoff of 0.001 (Hint: the summary method for GOHyperGResult will help a lot.) Is there a pattern?

Exercise 8.11
Use the GOTERM *annotation object to retrieve a more comprehensive description of some of the categories.*

8.5   Other annotations available

The Bioconductor annotation packages also contain mappings between identifiers and PubMed IDs. These can be used together with functions such as pm.getabst and pm.abstGrep to automatically download and search relevant PubMed abstracts for key words. Every annotation package also supplies mappings to PFAM and Prosite identifiers.

8.6   biomaRt

In the previous sections we have used annotation information that is available in Bioconductor through static annotation packages. There is a lot of additional information available in the numerous biological databases such as Ensembl or Uniprot (Flicek et al., 2007; UniProt, 2007). We can access these databases online, in R, using the package **biomaRt**. Here, we want to use **biomaRt** to look up 3ʹ UTRs of our differentially expressed genes. This information could for instance be used in the subsequent analysis of regulatory sequences such as microRNA binding sites. We first need to create an instance of the *Mart* class which stores the connection information to the database. All available BioMart Web services can be listed using the function `listMarts`. The function `head` reduces the output to the first couple of entries.

```
> library("biomaRt")
> head(listMarts())
                             name
1                         ensembl
2      compara_mart_homology_48
3 compara_mart_pairwise_ga_48
4 compara_mart_multiple_ga_48
5                             snp
6               genomic_features
                                  version
1                 ENSEMBL 48 GENES (SANGER)
2               ENSEMBL 48 HOMOLOGY (SANGER)
3 ENSEMBL 48 PAIRWISE ALIGNMENTS (SANGER)
4 ENSEMBL 48 MULTIPLE ALIGNMENTS (SANGER)
5               ENSEMBL 48 VARIATION (SANGER)
6     ENSEMBL 48 GENOMIC FEATURES (SANGER)
```

We use Ensembl for our example.

```
> mart = useMart("ensembl")
```

Often BioMart databases contain more than one dataset. We can check for available datasets using the function `listDatasets`.

```
> head(listDatasets(mart))
                          dataset
1          oanatinus_gene_ensembl
2         gaculeatus_gene_ensembl
3         cporcellus_gene_ensembl
4         lafricana_gene_ensembl
5 stridecemlineatus_gene_ensembl
```

```
6          scerevisiae_gene_ensembl
                                      description     version
1             Ornithorhynchus anatinus genes (OANA5)      OANA5
2             Gasterosteus aculeatus genes (BROADS1)     BROADS1
3                Cavia porcellus genes (GUINEAPIG) GUINEAPIG
4                Loxodonta africana genes (BROADE1)     BROADE1
5 Spermophilus tridecemlineatus genes (SQUIRREL)    SQUIRREL
6          Saccharomyces cerevisiae genes (SGD1.01)     SGD1.01
```

Our data are from human arrays, so we want to work with the *hsapiens_gene_ensembl* set, and need to update our *Mart* object accordingly.

```
> ensembl = useDataset("hsapiens_gene_ensembl",
      mart=mart)
```

For the Ensembl database **biomaRt** offers a set of convenience functions for the most common tasks. The function getGene uses a vector of query IDs to look up names, descriptions, and chromosomal locations of corresponding genes. getGo can be used to fetch GO annotations and getSequences retrieves different kinds of sequence information. getSNP and getHomolog are useful to query SNP data or to map gene identifiers from one species to another.

Exercise 8.12
Fetch the sequences of 3′ UTRs of our set of differentially expressed genes using getSequence. Take a look at its manual page to learn about the function's parameters. Think about which type of gene IDs we have available for our set of genes.

biomaRt allows us to retrieve many different kinds of data in a very flexible manner. To understand how its generalized query API works, we first have to learn about the terms *filter* and *attribute*. A filter defines the restriction on a query, for example, to show results only for a subset of genes selected by a gene identifier. Attributes define the values we want to retrieve, for instance, the IDs of PFAM domains for these genes. You can get a list of available filters with listFilters

```
> head(listFilters(ensembl, group="GENE:"))
                  name                 description
1         affy_hc_g110         Affy hc g 110 ID(s)
2       affy_hc_g110-2         Affy hc g 110 ID(s)
3        affy_hg_focus         Affy hg focus ID(s)
4      affy_hg_focus-2         Affy hg focus ID(s)
5   affy_hg_u133_plus_2 Affy hg u133 plus 2 ID(s)
6 affy_hg_u133_plus_2-2 Affy hg u133 plus 2 ID(s)
```

and of available attributes with `listAttributes`.

```
> head(listAttributes(ensembl, group="PROTEIN:"))
                          name                  description
1                       family          Ensembl Family ID
2           family_description          Family Description
3                     interpro                 Interpro ID
4         interpro_description        Interpro Description
5 interpro_short_description Interpro Short Description
6                         pfam                     PFAM ID
```

For some BioMart databases, in particular for Ensembl, there are many attributes and filters available, and you can control the number that are listed by the above functions with the `group` parameter. The general-purpose query interface of **biomaRt** is provided by the function `getBM`.

Exercise 8.13
For our set of differentially expressed genes, find associated protein domains. Such domains are stored for instance in the PFAM, Prosite, or InterPro databases. Try to find domain IDs for one or for all of these sources.

8.7   Database versions of annotation packages

As of release 2.1 of Biconductor, the environment-based annotation packages are being phased out in favor of a new set of packages where the data are stored in SQLite databases. Database variants all have the `.db` suffix, so that the database variant of the **hgu95av2** annotation package is called **hgu95av2.db**. Every effort has been made to ensure that the database variants can be used in essentially the same way as the environment-based packages. However, by using a relational database API many operations that used to be difficult are now much simpler.

The basic interface to these new packages is contained in the **AnnotationDbi** package, and its vignette should be consulted for details. Every db annotation package exports some functions that can be used to access the database directly. The names are mangled so that every package has a different name for these. To get a connection to the database, the function for the **hgu133a.db** package is `hgu133a_dbconn`.

```
> library("hgu133a.db")
> dbc = hgu133a_dbconn()
```

Almost all of the interactions that were supported for the environment-based annotation packages are also supported for the `db` versions. For example, you can access the data directly using any of the standard subsetting or extraction functions that also work for environments: `get`, `mget`, `$`, and `[[`.

```
> get("201473_at", hgu133aSYMBOL)
[1] "JUNB"
> mget(c("201473_at","201476_s_at"), hgu133aSYMBOL)
$`201473_at`
[1] "JUNB"

$`201476_s_at`
[1] "RRM1"
> hgu133aSYMBOL$"201473_at"
[1] "JUNB"
> hgu133aSYMBOL[["201473_at"]]
[1] "JUNB"
```

Some of the advantages to the new paradigm can be seen in the next example, where we construct a table counting the number of terms in each of the GO categories. The code using environment-based packages follows. (It will work for `db` packages as well.)

```
> goCats = unlist(eapply(GOTERM, Ontology))
> gCnums = table(goCats)[c("BP","CC", "MF")]
```

We can use the **xtable** package to display the result; see Table 8.1.

```
> library("xtable")
> xtable(as.matrix(gCnums), display=c("d", "d"),
      caption="Number of GO terms per ontology.",
      label="ta:GOprops")
```

The code below uses the new `db` functionality. It is much faster.

```
> query = "select ontology from go_term"
> goCats = dbGetQuery(GO_dbconn(), query)
> gCnums2 = table(goCats)[c("BP","CC", "MF")]
> identical(gCnums, gCnums2)
[1] TRUE
```

Table 8.1. Number of GO terms per ontology.

	x
BP	13916
CC	2007
MF	7878

We can search for GO terms containing the word `chromosome`. We first construct a SQL query, and then apply it.

```
> query = paste("select term from go_term where term",
      "like '%chromosome%'")
> chrTerms = dbGetQuery(GO_dbconn(), query)
> nrow(chrTerms)
[1] 82
> head(chrTerms)
                                         term
1                          nuclear chromosome
2                      cytoplasmic chromosome
3                     mitochondrial chromosome
4            chromosome, pericentric region
5           condensed chromosome kinetochore
6 condensed nuclear chromosome kinetochore
```

Next we show how to find the GO identifier for "transcription factor binding" and use that to get all Entrez Gene IDs with that annotation. Here, we restrict our attention to the genes covered by the HG-U133A GeneChip. Alternatively, you could use the `org.Hs.eg.db` package, which annotates the whole human genome.

```
> query = paste("select go_id from go_term where",
      "term = 'transcription factor binding'")
> tfb = dbGetQuery(GO_dbconn(), query)
> tfbps =  hgu133aGO2ALLPROBES[[tfb$go_id]]
> table(names(tfbps))
IDA IEA IEP IGI IMP IPI ISS NAS  NR TAS
106  83   2   2   8  59  30  91  27 366
```

Exercise 8.14
How many GO terms have the words `transcription factor` in them?

8.7.1 Mapping Symbols

In this section we address a more advanced topic. One of the problems with the environment-based system was that the probe ID was used as the primary key in most mappings which made it quite difficult to map between entities where neither was the primary key. With the database-based system that is much more straightforward. Let us consider a common problem, mapping from gene symbols to some other form of identifier. This problem arises when publications give the symbol (or name) of a gene, but not a systematic database identifier.

The code in the next segment consists of four functions, three helpers and the main function findEGs that maps from symbols to Entrez Gene IDs. We need to know about the table structure to write the helper functions as they are basically R wrappers around SQL statements. The hgu95av2_dbschema function can be used to obtain all the information about the schema.

```
> queryAlias = function(x) {
    it = paste("('", paste(x, collapse="', '"), "'", sep="")
    paste("select _id, alias_symbol from alias",
        "where alias_symbol in", it, ");")
}
> queryGeneinfo = function(x) {
    it = paste("('", paste(x, collapse="', '"), "'", sep="")
    paste("select _id, symbol from gene_info where",
        "symbol in", it, ");")
}
> queryGenes = function(x) {
    it = paste("('", paste(x, collapse="', '"), "'", sep="")
    paste("select * from genes where _id in", it,  ");")
}
> findEGs = function(dbcon, symbols) {
    rs = dbSendQuery(dbcon, queryGeneinfo(symbols))
    a1 = fetch(rs, n=-1)
    stillLeft = setdiff(symbols, a1[,2])

    if( length(stillLeft)>0 ) {
      rs = dbSendQuery(dbcon, queryAlias(stillLeft))
      a2 = fetch(rs, n=-1)
      names(a2) = names(a1)
      a1 = rbind(a1, a2)
    }

    rs = dbSendQuery(dbcon, queryGenes(a1[,1]))
    merge(a1, fetch(rs, n=-1))
}
```

The logic is to first look to see if the symbol is currently in use, and then for those that were not found to search in the alias table, to see if there are updated names. Each of the first two queries within the findEGs function returns the symbol (the second columns of a1 and a2) and an identififer that is internal to the SQLite database (the first columns). The last query uses those internal IDs to extract the corresponding EntrezGene IDs.

```
> findEGs(dbc, c("ALL1", "AF4", "BCR", "ABL"))
  _id symbol gene_id
1  20    ABL      25
2 543    BCR     613
3 3781   AF4    4299
4 3946   ABL    4547
```

The three columns in the return are the internal ID, the symbol, and the EntrezGene ID (`gene_id`).

8.7.2 Other capabilities

In many situations you may want to reverse a mapping. That is, you have an annotation that goes from Affymetrix ID to symbol, and you would like to have the mapping from symbols to Affymetrix IDs. This is easily done using the `revmap` function.

```
> s1 = revmap(hgu133aSYMBOL)
> s1$BCR
[1] "202315_s_at" "214623_at"   "217223_s_at"
```

Another useful tool in the **AnnotationDbi** package is the `toTable` function. It takes as input an instance of the *Bimap* class and returns a *data.frame*. You may want to consult the documentation of the **AnnotationDbi** package for more details. In the code below we show how to obtain GO information using `toTable`.

```
> toTable(hgu133aGO["201473_at"])
     probe_id        go_id Evidence Ontology
1  201473_at GO:0000074      IEA       BP
2  201473_at GO:0006357      TAS       BP
3  201473_at GO:0009987      IEA       BP
4  201473_at GO:0000785      TAS       CC
5  201473_at GO:0005634      IEA       CC
6  201473_at GO:0003700      IEA       MF
7  201473_at GO:0003702      TAS       MF
8  201473_at GO:0003713      TAS       MF
9  201473_at GO:0003714      TAS       MF
10 201473_at GO:0043565      IEA       MF
11 201473_at GO:0046983      IEA       MF
```

9

Supervised Machine Learning

R. Gentleman, W. Huber, and V. J. Carey

Abstract

In this chapter we cover some of the basic principles of supervised machine learning. We mainly consider the two-class problem, but also cover some multiclass prediction. We introduce some of the basic concepts in machine learning such as the distance function, the so-called *confusion matrix*, and cross-validation. We make extensive use of the **MLInterfaces** package.

9.1 Introduction

Machine learning (ML) is typically divided into two separate areas, supervised ML and unsupervised ML. The first of these is referred to as classification in the statistics literature, and the second is referred to as clustering. Both types of machine learning are concerned with the analysis of datasets containing multivariate observations. There is a large amount of literature that can provide an introduction into these topics; here we refer to Breiman et al. (1984) and Hastie et al. (2001).

In supervised learning, a p-dimensional multivariate observation x is associated with a class label c. The p components of datum x are called features. The objective is to "learn" a mathematical function f that can be evaluated on the input x to yield a prediction of its class c. We consider the case in which a training set of multivariate observations and associated class labels is provided. We introduce software in Bioconductor that makes it easy to estimate prediction functions f on the basis of a training set, and to compute predictions using a distinct test set, for which class labels are not available. We also consider methods for assessing the likely error rate of the resulting classifiers.

One issue that typically arises in ML applications to high-throughput biological data is feature selection. For example, in the case of microarray

F. Hahne et al., *Bioconductor Case Studies*, DOI: 10.1007/978-0-387-77240-0_9,
© Springer Science+Business Media, LLC 2008

data one typically has tens of thousands of features that were collected on all samples, but many will correspond to genes that are not expressed. Other features will be important for predicting one phenotype, but largely irrelevant for predicting other phenotypes. Thus, feature selection is an important issue.

Fundamental to the task of ML is selecting a measure of similarity among (or distance between) multivariate data points. We emphasize the term "selecting" here because it can easily be forgotten that the units in which features have been measured have no legitimate priority over other transformed representations that may lead to more biologically sensible criteria for classification. If we simply drop our expression data into a classification procedure, we have made an implicit selection to embed our observations in the feature space employed by the procedure. Oftentimes this feature space has Euclidean structure. If we extended our expression data to include, say, squares of expression values for certain genes, or products of values taken on several genes, a given classification procedure may perform very differently, even though the original data have only been deterministically transformed. Effective classification requires attention to the possible transformations (equivalently, distance metric in the implied feature space) of complex machine learning tools such as kernel support vector machines. In many cases the distance metric is more important than the choice of classification algorithm, and **MLInterfaces** makes it reasonably easy to explore different choices for distances. In this chapter we concentrate on the problem of classifying samples, but the methods can also be applied to classifying features (genes, if we are using expression microarrays).

9.1.1 Supervised machine learning check list

1. Filter out features (genes) that show little variation across samples, or that are known not to be of interest. If appropriate, transform the data of each feature so that they are all on the same scale.

2. Select a distance, or similarity, measure. What does it mean for two samples to be close? Make sure that the selected distance embodies your notion of similarity.

3. Feature selection: Select features to be used for ML. If you are using cross-validation, be sure that feature selection according to your criteria, which may be data-dependent, is performed at each iteration.

4. Select the algorithm: Which of the many ML algorithms do you want to use?

5. Assess the performance of your analysis. With supervised ML, performance is often assessed using cross-validation, but this itself can be performed in various ways.

9.2   The example dataset

For this chapter we need a dataset that allows for two, or more, group comparisons. The ALL dataset contains over 100 samples, for a variety of different subtypes of the disease. In the code below we load the data, and then subset to the particular phenotypes in which we are interested. The specific information we need is to select those with B-cell ALL, and then within that subset, those that are NEG and those that are labeled as BCR/ABL. The last line in the code below is used to drop unused levels of the factor encoding `mol.biol`.

```
> library("ALL")
> data(ALL)
> bcell = grep("^B", as.character(ALL$BT))
> moltyp = which(as.character(ALL$mol.biol)
      %in% c("NEG", "BCR/ABL"))
> ALL_bcrneg = ALL[, intersect(bcell, moltyp)]
> ALL_bcrneg$mol.biol = factor(ALL_bcrneg$mol.biol)
```

Exercise 9.1
How many samples for each class are in the BCR/ABL-NEG subset?

The comparison of BCR/ABL to NEG is difficult, and the error rates are typically quite high. You could instead compare BCR/ABL to ALL1/AF4; they are rather easy to distinguish and the error rates should be smaller.

9.2.1   Nonspecific filtering of features

Nonspecific filtering removes those genes that we believe are not sufficiently informative for any phenotype, so that there is little point in considering them further. For the purpose of this teaching exercise, we used a very stringent filter so that the dataset is small and the examples will run quickly; in practice you would probably use a less stringent filter.

We use the function `nsFilter` from the **genefilter** package to filter for a number of different criteria. For instance, by default it removes the control probes on Affymetrix arrays, which can be identified by their **AFFX** prefix. We also exclude genes without Entrez Gene identifiers, and as suggested above, we select the top 25% of genes on the basis of variability across samples.

```
> ALLfilt_bcrneg = nsFilter(ALL_bcrneg, var.cutoff=0.75)$eset
```

Exercise 9.2
What kind of object is `ALLfilt_bcrneg`?

9.3   Feature selection and standardization

Feature selection is an important component of machine learning. Typically the identification and selection of features used for supervised ML relies on knowledge of the system being studied, and on univariate assessments of predictive capability. Among the more commonly used methods are the selection of features that are predictive using t-statistic and ROC curves (at least for two-sample problems).

In order to correctly assess error rates it is essential to accounted for the effects of feature selection. If cross-validation is used then feature selection must be incorporated within the cross-validation process and not performed ahead of time using all of the data.

A second important aspect is standardization. For gene expression data the recorded expression level is not directly interpretable, and so users must be careful to ensure that the statistics used are comparable. This standardization ensures that all genes have equal weighting in the ML applications. In most cases this is most easily achieved by standardizing the expression data, within genes, across samples. In some cases (such as with a t-test) there is no real need to standardize because the statistic itself is standardized.

In the code segments below we standardize all gene expression values. It is important that nonspecific filtering has already been performed. We first write a helper function to compute the rowwise IQRs for us.

```
> rowIQRs = function(eSet) {
     numSamp = ncol(eSet)
     lowQ = rowQ(eSet, floor(0.25 * numSamp))
     upQ = rowQ(eSet, ceiling(0.75 * numSamp))
     upQ - lowQ
  }
```

Next we subtract the row medians and divide by the row IQRs. Again, we write a helper function, standardize, that does most of the work.

```
> standardize = function(x) (x - rowMedians(x)) / rowIQRs(x)
> exprs(ALLfilt_bcrneg) = standardize(exprs(ALLfilt_bcrneg))
```

9.4   Selecting a distance

To some extent your choices here are not always that flexible because many ML algorithms have a certain choice of distance measure, say, the Euclidean distance, built in. In such cases, you still have the choice of transformation of the variables; examples are coordinatewise logarithmic transformation, the linear Mahalonobis transformation, and other linear or nonlinear projections of the original features into a (possibly lower-dimensional) space.

If the ML algorithm does allow explicit specification of the distance metric, there are a number of different tools in R to compute the distance between objects. They include the function dist, the function daisy from the **cluster** package (Kaufman and Rousseeuw, 1990), and the functions in the **bioDist** package. The **bioDist** package is discussed in Chapter 12 of Gentleman et al. (2005a). Some ideas on visualizing distance measures can be found in Chapter 10.5 of that reference.

Exercise 9.3
*What distance measures are available in the **bioDist** package? Hint: load the package and then look at the loaded functions, or read the vignette.*

The dist function computes the distance between rows of an input matrix. We want the distances between samples, thus we transpose the matrix using the function t. The return value is an instance of the *dist* class. Because this class is not supported by some R functions that we want to use, we also convert it to a matrix.

```
> eucD = dist(t(exprs(ALLfilt_bcrneg)))
> eucM = as.matrix(eucD)
> dim(eucM)
[1] 79 79
```

We next visualize the distances using a heatmap. In the code below we generate a range of colors to use in the heatmap. The **RColorBrewer** package provides a number of different palettes to use and we have selected one that uses red and blue. Because we want red to correspond to high values, and blue to low, we must reverse the palette.

```
> library("RColorBrewer")
> hmcol = colorRampPalette(brewer.pal(10, "RdBu"))(256)
> hmcol = rev(hmcol)
> heatmap(eucM, sym=TRUE, col=hmcol, distfun=as.dist)
```

The result of this is shown in Figure 9.1.

Exercise 9.4
What do you notice most about the heatmap? What color is used to encode objects that are similar? What color encodes objects that are dissimilar?

Exercise 9.5
Repeat this analysis using Spearman's correlation distance. How much does the heatmap change?

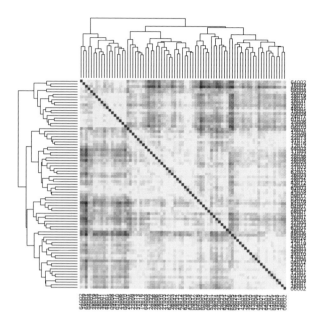

Figure 9.1. A heatmap of the between-sample distances.

For the next exercise, we use a helper function from the **bioDist** package, closest.top, that finds the nearest neighbors of a particular observation, given a distance matrix x.

```
> closest.top("03002", eucM, 1)
[1] "09017"
```

Exercise 9.6
*Compute the distance between the samples using the MIdist function from the **bioDist** package. What distance does this function compute? Which sample is closest to "03002" in this distance?*

9.5   Machine learning

There are many different ML algorithms available in R and through its many add-on packages. The user interfaces (i.e., the calling parameters and return values of the machine learning algorithms that are available in R) are quite diverse, and this can make switching your application code from one machine learning algorithm to another tedious. For this reason, the **MLInterfaces** provides *wrappers* around the various machine learning algorithms that accept a standardized set of calling parameters and produce a standardized return value. The package does not implement any of the machine learning algorithms, it just converts the in- and out-going data structures into the appropriate format. In general, the name of the function or method remains the same, but an I is appended, so we, for instance, use the **MLInterfaces** functions knnI to interface to the functions knn from the **class** package.

Exercise 9.7
Use `library(help=MLInterfaces)`, `?MLearn`, *and* `openVignette()` *to explore the package. Try to follow the example at the bottom of the* `MLearn` *help page.*

We start by looking at the *k* nearest neighbors (KNN) and diagonal linear discriminant analysis (DLDA) methods, because they are conceptually simple and serve well to demonstrate most of the general principles.

It is easiest to understand most supervised ML methods in the setting where one has both a training set on which to build the model, and a test set on which to test the model. We begin by artificially dividing our data into a test and training set. Such a dichotomy is not actually that useful and in practice one tends to rely on cross-validation, or other similar schemes.

```
> Negs = which(ALLfilt_bcrneg$mol.biol == "NEG")
> Bcr = which(ALLfilt_bcrneg$mol.biol == "BCR/ABL")
> S1 = sample(Negs, 20, replace=FALSE)
> S2 = sample(Bcr, 20, replace = FALSE)
> TrainInd = c(S1, S2)
> TestInd = setdiff(1:79, TrainInd)
```

The term *confusion matrix* is typically used to refer to the table that cross-classifies the test set predictions with the true test set class labels. The **MLInterfaces** packages provides a function called confuMat that will compute this matrix from most inputs.

In the next series of exercises you are introduced to the basic functions in the **MLInterfaces** package and through them to several supervised machine learning methods.

Exercise 9.8
Using the KNN, LDA, and DLDA methods predict the phenotype (BCR/ABL or NEG) for the samples in the `ALLfilt_bcrneg` *dataset. Estimate the prediction error rates. Later in Section 9.6 we show how to use cross-validation to estimate the prediction error rates.*

In every machine learning algorithm one can, at least conceptually, make one of three decisions:

1. To classify the sample into one of the known classes as defined by the training set.

2. To indicate doubt, the sample is somehow between two or more classes and there is no clear indication as to which class it belongs.

3. To indicate that the sample is an outlier, in the sense that it is so dissimilar to all samples in the training set that no sensible classification is possible.

In the next series of exercises we guide you through some of the options that are available. Unfortunately these concepts are not implemented in all (or even most) machine learning algorithms and hence their usage is problematic.

Exercise 9.9
For the KNN classifier, answer the following questions.

   a. *What happens when k is even and there is a tie?*

   b. *Optional: Suppose that instead of Euclidean distance you wanted to use some other metric, such as 1-correlation. How might you achieve that?*

   c. *How might you define outlier and doubt classes? Are there any outliers, or hard to classify samples?*

The preceding discussion and exercises used all of the features that passed our nonspecific filtering procedure. But some are not likely to be predictive of the phenotypes of interest, and so we now want to explore what happens if we instead select genes that are able to discriminate between those with BCR/ABL and those samples labeled NEG. We use the *t*-test to select genes; those with small *p*-values for comparing BCR/ABL to NEG are used. Although it is tempting to use all the data to do this selection, that is not really a good idea as it tends to give misleadingly low values for the error rates. You can, and probably should, use attenuated *t*-tests, and you can select the ones to use by the observed *p*-value. But, these approaches would complicate the exposition further, so we simply select those 50 genes with the most extreme *t*-statistics.

In the code below, we compute the *t*-tests on the training set, then sort them from largest to smallest, and then obtain the names of the 50 that have the largest observed test statistics.

```
> Traintt = rowttests(ALLfilt_bcrneg[, TrainInd], "mol.biol")
> ordTT = order(abs(Traintt$statistic), decreasing=TRUE)
> fNtt = featureNames(ALLfilt_bcrneg)[ordTT[1:50]]
```

Exercise 9.10
Repeat this exercise on the whole dataset. How many of the genes selected on the training set were also selected when you used the whole dataset?

Now we can see how well the different machine learning algorithms work when the features have been selected to help discriminate between the two groups. First we use KNN.

```
> BNf = ALLfilt_bcrneg[fNtt,]
> knnf = MLearn( mol.biol ~ ., data=BNf, knnI(k=1,l=0),
        TrainInd)
>  confuMat(knnf)
          predicted
given      BCR/ABL NEG
  BCR/ABL      15   2
  NEG           5  17
```

Exercise 9.11
Repeat this with LDA and DLDA and compare the error rates obtained for these three methods using selected genes with those obtained using all of the genes.

9.6   Cross-validation

Assessing the error rate in supervised machine learning is important, but potentially problematic. It is well known that the error rate is overly optimistic if the same data are used to estimate it as were used to train the model. This led to an approach that divided the data into two groups, as was done in the previous section, one for training the model and one for testing (or assessing the error rate). However, that approach is somewhat inefficient, and cross-validation is generally preferable as an approach.

Cross-validation is a very useful tool that can be applied to many different problems. It can be used for model selection, selecting parameter values for a given algorithm, and for assessing error rates in classification problems, to name but a few of the many areas to which it has been applied. The basic idea behind this method is quite simple: one must be willing to believe that the dataset one has can be divided into two pieces, and that for such a division it makes sense to fit a model to one piece, and assess the performance of that model on the other. And under such an assumption, there are typically many, nearly equivalent, ways to divide the data so rather than do this once, we should consider many different divisions. Then, for error rate assessment we fit our model to the training set, estimate the error rate on the test set, aggregate over all divisions, and thereby obtain an estimate of the error rate. In order to get an accurate assessment it is important that all steps that can affect the outcome are included in the cross-validation process. In particular, the selection of features to use in the machine learning algorithm must be included within the cross-validation step.

Perhaps the easiest method to understand, and the most widely used method, is leave-one-out (LOO) cross-validation. In this scheme, each observation is left out in turn, the remaining $n - 1$ observations are used as the training set, and the left-out observation is treated as the test set. There are many other ways to perform cross-validation, but all are more complicated than the LOO scheme and hence require more thought. It is also common to partition the data into tenths, and use one tenth as the test set, while using the other nine tenths as the training set. Although there are some esthetic reasons to use a partition one might also just use randomly selected subsets, even if there is some overlap. This approach has the benefit that there are in fact many more such subsets than partitions, and hence one might obtain a better estimate of the mean error rate.

The **MLInterfaces** package has a mechanism for performing cross-validation. The mechanism is based on specifying an xvalSpec parameter to the MLearn function. The xvalSpec allows you to specify a type (if "LOO" then the other parameters are ignored), a partition function (for specifying the test and training sets), the number of partitions, and optionally a function that helps to select features in each subset.

Because cross-validation is a very expensive operation, and these exercises are intended to run on laptop computers, we first artificially reduce the size of the dataset, to 1000 genes for the remainder of this section.

```
> BNx = ALLfilt_bcrneg[1:1000,]
```

The example below performs LOO cross-validation, using KNN. This is a bit special, because the **class** provides a purpose-built function for LOO cross-validation using KNN and we want to access it directly. The one, slightly odd requirement is to specify that all samples are part of the training set.

```
> knnXval1 = MLearn(mol.biol~., data=BNx, knn.cvI(k=1, l=0),
        trainInd=1:ncol(BNx))
```

In the code below, we show how you could perform essentially the same analysis using the xvalSpec approach. This is a much more flexible approach, but unfortunately it is less efficient especially with large datasets, such as we are using. The first two arguments should be familiar. The third argument, knnI, specifies that we use the knn function. The final argument, xvalSpec, indicates the method that will be used for cross-validation and LOO stands for leave-one-out.

```
> knnXval1 = MLearn(mol.biol~., data=BNx, knnI(k=1, l=0),
      xvalSpec("LOO"))
```

Exercise 9.12

  a. *Describe in words the operation that the code is performing.*

  b. *What information is provided by the confuMat function? How would you use this to assess the performance of this machine learning algorithm?*

```
> confuMat(knnXval1)
        predicted
given     BCR/ABL NEG
  BCR/ABL     31   6
  NEG         16  26
```

Now, let us see what happens when we include feature selection in the cross-validation. This can be done by invoking a helper function, fs.absT, as part of xvalSpec. In order to include features that produce the top N two-sample t-statistics (in absolute value) among all genes, pass fs.absT(N) as the fourth argument to xvalSpec:

```
> lk3f1 = MLearn(mol.biol~., data=BNx, knnI(k=1),
      xvalSpec("LOO", fsFun=fs.absT(50)))
> confuMat(lk3f1)
```

Exercise 9.13

  a. *In the example above we used 50 features for each of the cross-validations. What happens if we only use 5? How would you interpret these results? Which features were selected, and how many times, when we used 5 features (Hint: fsHistory)?*

  b. *Optional: Hard. Repeat the exercise above using tenfold cross-validation. To do this you need to divide the data into ten groups and use the group argument to xval.*

Cross-validation can be used for many different purposes. In the next series of exercises we guide you through the use of cross-validation for selection of the parameter k in KNN.

Exercise 9.14
Use cross-validation to estimate k, the number of nearest neighbors to use. That is, for each of a number of values of k, estimate the cross-validation error, and then select k as that value which yields the smallest error rate.

9.7 Random forests

In this section we describe random forests (Breiman, 1999) and provide some examples and exercises based on the **randomForest** package (Liaw and Wiener, 2002), but use the **MLInterfaces** interface. Basic use of random forest technology is fairly straightforward. The only parameter that seems to be very important is `mtry`. This controls the number of features that are selected for each split. The default value is the square root of the number of features but often a smaller value tends to have better performance. In the code below we fit two *forests* with quite different values of `mtry` to help see what effect that might have. The seed for the random number generator is set to ensure repeatability.

It is not typical to produce a test and separate training set, as we have done here, when using random forests. We use the `MLearn` interface, and request that the different measures of variable importance be retained (they are explained below).

```
> library("randomForest")
> set.seed(123)
> rf1 = MLearn( mol.biol~., data=ALLfilt_bcrneg,
      randomForestI, TrainInd, ntree=1000, mtry=55,
      importance=TRUE)
```

Next we use a much smaller value of `mtry` so that we can compare the results.

```
> rf2 = MLearn( mol.biol~., data=ALLfilt_bcrneg,
      randomForestI, TrainInd, ntree=1000, mtry=10,
      importance=TRUE)
```

We can use the prediction function to assess the ability of these two forests to predict the class for the test set. For each model we show the confusion matrix for both the training and test sets. Naturally the error rates are much smaller (zero in both cases) for the training set.

```
> trainY = ALLfilt_bcrneg$mol.biol[TrainInd]
> confuMat(rf1, "train")
        predicted
given     BCR/ABL NEG
  BCR/ABL     20   0
  NEG          0  20
> confuMat(rf1, "test")
        predicted
given     BCR/ABL NEG
  BCR/ABL     16   1
  NEG          4  18
```

Now for the second model.

```
> confuMat(rf2, "train")
        predicted
given     BCR/ABL NEG
  BCR/ABL     20   0
  NEG          0  20
> confuMat(rf2, "test")
        predicted
given     BCR/ABL NEG
  BCR/ABL     14   3
  NEG          4  18
```

Exercise 9.15
*Compare the error rates from the two different random forest fits. Compare
the error rates from the random forest fits to those for KNN.*

9.7.1   Feature selection

One of the nice things about the random forest technology is that it provides
an indication of which variables were most important in the classification
process. The specific definitions of these measures are provided in the manual page for the `importance` function, which can be used to extract the
measures from an instance of the *randomForest* class. These features can
be compared to those selected by *t*-test or selected by some other means.
In the next two code chunks we plot the variable importance statistics for
the two random forests. The output is shown in Figure 9.2.

```
> opar = par(no.readonly=TRUE, mar=c(7,5,4,2))
> par(las=2)
> impV1 = getVarImp(rf1)
> plot(impV1, n=15, plat="hgu95av2", toktype="SYMBOL")
> par(opar)
```

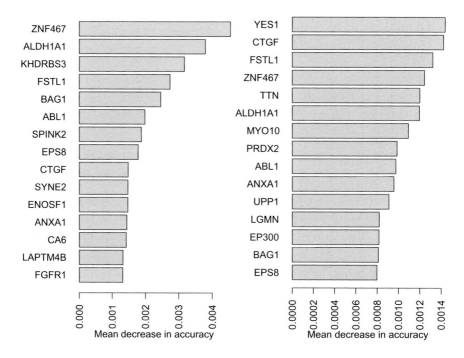

Figure 9.2. Variable importance plots: the left panel is for the first random forest model (`mtry = 55`), the right panel is for the second model (`mtry = 10`).

```
> par(las=2, mar=c(7,5,4,2))
> impV2 = getVarImp(rf2)
> plot(impV2, n=15, plat="hgu95av2", toktype="SYMBOL")
> par(opar)
```

Exercise 9.16
Are there any variables in common among the 15 most important for each model?

9.7.2   More exercises

Again a number of interesting exercises present themselves.

Exercise 9.17
  a. *Reverse the role of the test set and the training set and see how the estimated prediction errors change.*

  b. *Use the whole dataset to build a random forest. How well does it do?*

A minor caveat to the use of random forests is that the method seems to have difficulties when the sizes of the groups are not approximately equal. There is a `weight` argument that can be given to the random forest function but it appears to have little effect.

9.8   Multigroup classification

We now briefly consider the application of supervised machine learning methods to a multiclass problem. We return to our original data, and instead of considering a two-class problem, we consider three different classes, BCR/ABL, NEG and ALL1/AF4. The code below creates an expression set containing these three groups. We perform some nonspecific filtering, and the rather cryptic last line of the code chunk drops unused levels of the factor `mol.biol`. Please see Chapter 8 for a discussion on the choice of varinance measure.

```
> Bcell = grep("^B", ALL$BT)
> ALLs = ALL[,Bcell]
> types = c("BCR/ABL", "NEG", "ALL1/AF4")
> threeG = ALLs$mol.biol %in% types
> ALL3g = ALLs[,threeG]
> qrange <- function(x)
        diff(quantile(x, c(0.1, 0.9)))
> ALL3gf = nsFilter(ALL3g, var.cutoff=0.75,
        var.func=qrange)$eset
> ALL3gf$mol.biol = factor(ALL3gf$mol.biol)
```

We artificially divide the data set into test and training sets, so that a model can be built on the training set and tested on the test set. Because the different subtypes have very different sizes, we attempt to balance our selection.

```
> s1 = table(ALL3gf$mol.biol)
> trainN = ceiling(s1/2)
> sN = split(1:length(ALL3gf$mol.biol), ALL3gf$mol.biol)
> trainInd = NULL
> testInd = NULL
> set.seed(777)
> for(i in 1:3) {
        trI = sample(sN[[i]], trainN[[i]])
        teI = setdiff(sN[[i]], trI)
        trainInd = c(trainInd, trI)
        testInd = c(testInd, teI)
  }
```

```
> trainSet = ALL3gf[, trainInd]
> testSet = ALL3gf[, testInd]
```

Exercise 9.18

Now use the KNN procedure to make class predictions. Can you estimate the class-conditional error rates? Can you control the procedure so that the class-conditional error rates are treated equally?

Exercise 9.19

Repeat the above classification using random forests.

10

Unsupervised Machine Learning

R. Gentleman and V. J. Carey

Abstract

In this chapter we explore the use of unsupervised machine learning, or clustering. We cover distances, dimension reduction techniques, and a variety of unsupervised machine learning methods including hierarchical clustering, k-means clustering, and specialized methods, such as those in the **hopach** package.

10.1 Preliminaries

Cluster analysis is also known as unsupervised machine learning, and has a long and extensive history. There are many good references that cover some of the topics discussed here in more detail, such as Gordon (1999), Kaufman and Rousseeuw (1990), Ripley (1996), Venables and Ripley (2002), and Pollard and van der Laan (2005). Unsupervised machine learning is also sometimes referred to as class discovery. One of the major differences between unsupervised machine learning and supervised machine learning is that there is no training set for the former and hence, no obvious role for cross-validation. A second important difference is that although most clustering algorithms are phrased in terms of an optimality criterion there is typically no guarantee that the globally optimal solution has been obtained. The reason for this is that typically one must consider all partitions of the data, and for even moderate sample sizes this is not possible, so some heuristic approach is taken. Thus we recommend that where possible you should use different starting parameters.

The prerequisites to performing unsupervised machine learning are the selection of samples, or items to cluster, the selection of features to be used in the clustering, the choice of similarity metric for the comparison of samples, and the choice of an algorithm to use. In this chapter we consider

F. Hahne et al., *Bioconductor Case Studies*, DOI: 10.1007/978-0-387-77240-0_10,
© Springer Science+Business Media, LLC 2008

the problem of clustering samples, but most of the methods would apply equally well to the problem of clustering genes.

There are two basic clustering strategies: `hierarchical clustering` and 2) `partitioning`, as well as some hybrid methods. Hierarchical clustering can be further divided into two flavors, *agglomerative* and *divisive*. In agglomerative clustering, each object starts as its own single-element cluster and at each stage the two closest clusters are combined into a new, bigger cluster. This procedure is iterated until all objects are in one cluster. The result of this process is a tree, which is often plotted as a dendrogram (see Figure 10.3). To obtain a clustering with a desired number of clusters, one simply cuts the dendrogram at the desired height. On the other hand, divisive hierarchical clustering begins with all objects in a single cluster. At each step of the iteration, the most heterogeneous cluster is divided into two, and this process is repeated until all objects are in their own cluster. The result is again a tree.

Partitioning algorithms typically require the number of clusters to be specified in advance. Then, samples are assigned to clusters, in some fashion, and a series of iterations, where (typically) single sample exchanges or moves are proposed and the resulting change in some clustering criteria computed; changes that improve the criteria are accepted. The process is repeated until either nothing changes or some number of iterations is made.

10.1.1   Data

First we load the necessary packages and load the dataset we use for the examples and exercises.

We use the ALL dataset, from the **ALL** package for this chapter. It is described more completely in Chapter 1. Our goal is to demonstrate how one can use various clustering methods, so we ignore the sample data. We reduce the data to a manageable size by selecting those samples that correspond to B-cell ALL and where the molecular biology phenotype is either BCR/ABL or NEG. The code for selecting the appropriate subset is given below; more details on the steps involved are given in Chapter 1.

```
> library("ALL")
> data(ALL)
> bcell = grep("^B", as.character(ALL$BT))
> moltyp = which(as.character(ALL$mol.biol)
      %in% c("NEG", "BCR/ABL"))
> ALL_bcrneg = ALL[, intersect(bcell, moltyp)]
> ALL_bcrneg$mol.biol = factor(ALL_bcrneg$mol.biol)
> ALLfilt_bcrneg = nsFilter(ALL_bcrneg, var.cutoff=0.75)$eset
```

The filtering has selected 2638 genes that we consider of interest for further investigation. This will still be too many genes for most applications

Table 10.1. GO molecular function categories that correspond to transcription factors.

GO Identifier	Description
GO:0003700	Transcription factor activity
GO:0003702	RNA polymerase II transcription factor activity
GO:0003709	RNA polymerase III transcription factor activity
GO:0016563	Transcriptional activator activity
GO:0016564	Transcriptional repressor activity

and often one will want to use other criteria to further reduce the genes under study. Here, we focus on transcription factors; these are important regulators of gene expression. As it turns out, finding the set of known transcription factors for any species is not such an easy problem. We use the GO identifiers in Table 10.1 that were used by Kummerfeld and Teichmann (2006) as their reference set of known transcription factors.

For each annotation of a gene to a GO category, there is an evidence code that indicates the basis for mapping that gene to the category. We drop all those that correspond to IEA, which stands for inferred from electronic annotation. We show the code for this task below.

```
> GOTFfun = function(GOID) {
      x = hgu95av2GO2ALLPROBES[[GOID]]
      unique(x[ names(x) != "IEA"])
  }
> GOIDs = c("GO:0003700", "GO:0003702", "GO:0003709",
      "GO:0016563", "GO:0016564")
> TFs = unique(unlist(lapply(GOIDs, GOTFfun)))
> inSel = match(TFs, featureNames(ALLfilt_bcrneg), nomatch=0)
> es2 = ALLfilt_bcrneg[inSel,]
```

This leaves us with 249 transcription factor coding genes for our machine learning exercises.

10.2 Distances

As we noted in the other machine learning exercise, no machine learning can take place without some notion of distance. It is not possible to cluster or classify samples without some way to say what it means for two things to be similar. For this reason, we again begin by considering distances. The dist function in R, the **bioDist** package, and the function daisy in the **cluster** package all provide different distances that you can use. It is always worth spending some time considering what it means for two objects to be similar and to then select a distance measure that reflects your belief.

Many machine learning methods have a built-in distance, often not obvious and difficult to alter, and if you want to use those methods you may need to use their metric. But it is important to realize that if you do use different measures of distance, they will have an impact on your analysis.

We begin by making use of the Manhattan metric; you might choose a different metric to compute distances between samples. Because we have no a priori belief that any one gene is more important than any other, we first center and scale the gene expression values before computing distances. Finally, we produce a heatmap based on the computed between-sample distances (Figure 10.1). There are no obvious groupings of samples based on this heatmap. We choose colors for our heatmap from a palette in the **RColorBrewer** package. Because the palette goes from red to blue, but we want high values to be red, we must reverse the palette, as is done in the code below.

```
> iqrs = esApply(es2, 1, IQR)
> gvals = scale(t(exprs(es2)), rowMedians(es2),
      iqrs[featureNames(es2)])
> manDist = dist(gvals, method="manhattan")
> hmcol = colorRampPalette(brewer.pal(10, "RdBu"))(256)
```

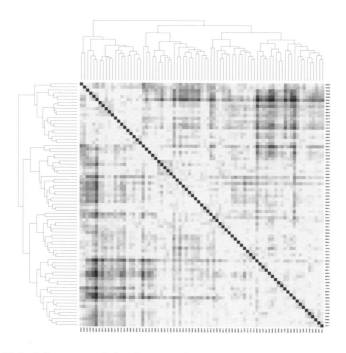

Figure 10.1. A heatmap of the distances between samples. Blue corresponds to small distances, red to large.

```
> hmcol = rev(hmcol)
> heatmap(as.matrix(manDist), sym=TRUE,  col=hmcol,
     distfun=function(x) as.dist(x))
```

Another popular visualization method for distance matrices is to use multidimensional scaling to reduce the dimensionality to two or three, and to then plot the resulting data. There are several different methods available, from the classical `cmdscale` function to Sammon mapping via the `sammon` function in the **MASS** package. Again we see little evidence of any grouping of the samples (Figure 10.2).

```
> cols = ifelse(es2$mol.biol == "BCR/ABL", "black",
     "goldenrod")
> sam1 = sammon(manDist, trace=FALSE)
> plot(sam1$points, col=cols, xlab="Dimension 1",
     ylab="Dimension 2")
```

Exercise 10.1
  a. *In the code above we obtained a two-dimensional reduction. Obtain a three-dimensional reduction, and if you have it installed, view this using the **rgl** package, so that you can rotate the points in three dimensions.*

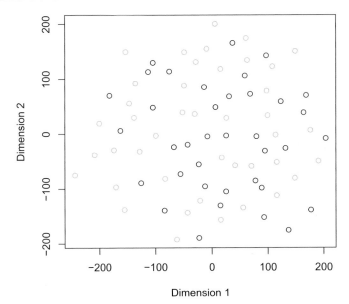

Figure 10.2. The two-dimensional projection of the between-sample distances obtained using Sammon mapping. Samples with the BCR/ABL phenotype are indicated with dark circles, and those with the NEG phenotype by light circles.

   b. *Repeat the example using classical multidimensional scaling.*

   c. *Repeat the exercise, but first restrict the genes you use to the 50 genes that best separate the two groups via their t-statistics.*

10.3 How many clusters?

Now that we have decided on a distance to use, we can ask one of the more fundamental questions that arises in any application of unsupervised machine learning: How many clusters are there? And unfortunately, even after a lot of research there is no definitive answer. The references given above provide some methods, and there are newer results as well, but none has been found to be broadly useful. We recommend visualizing your data, as much as possible, for instance, by using dimension reduction methods such as multidimensional scaling, as well as special-purpose tools such as the silhouette plot of Kaufman and Rousseeuw (1990); (see Section 10.8).

    Another popular method is to examine the dendrogram that is produced by some hierarchical clustering algorithm to see if it suggests a particular number of clusters. Unfortunately, this procedure is not really a good idea. If you compare the four dendrograms in Figure 10.3 they do not convey a coherent message. The third from the top suggests that there might be three clusters, but the other three are much less suggestive.

    The **hopach** package contains two functions that can be used to estimate the number of clusters. They are based on approaches that are related to the silhouette plot that is described in Section 10.8. In the code chunk below we demonstrate their use, on both the samples and the genes from our example dataset.

```
> mD = as.matrix(manDist)
> silEst = silcheck(mD, diss=TRUE)
> silEst
[1] 2.000 0.163
> mssEst = msscheck(mD)
> mssEst
[1] 4.0000 0.0777
> d2 = as.matrix(dist(t(gvals), method="man"))
> silEstG = silcheck(d2, diss=TRUE)
> silEstG
[1] 3.000 0.107
> mssEstG = msscheck(d2)
> mssEstG
[1] 6.0000 0.0489
```

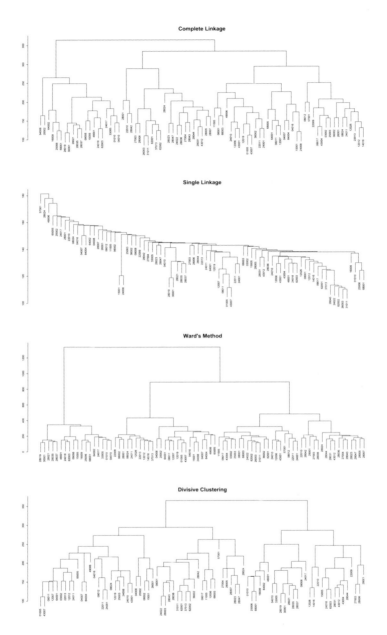

Figure 10.3. Four different dendrograms, clustering samples. The first three (top to bottom) were computed using agglomerative hierarchical clustering with different linkage methods, the bottom one used divisive hierarchical clustering.

The `silcheck` function returns a vector of length two; the first element is the recommended number of clusters, whereas the second element is the average silhouette for that number of clusters. The return value of `msscheck` is also of length two, the first value again being the recommended number of clusters, and for this function the second value is the median split silhouette (MSS).

We can see that `silcheck` recommends two, whereas `mssEst` recommends four. If instead, we consider clustering genes then the two methods recommend three and six clusters, respectively. These estimates could be used when we consider the partitioning methods described in Section 10.5.

Exercise 10.2
Repeat the exercise of assessing how many clusters there are, using another distance measure.

10.4   Hierarchical clustering

We now briefly discuss different hierarchical clustering methods. There are two basic strategies that can be used in hierarchical clustering. Divisive clustering begins with all objects in one cluster and at each step splits one cluster to increase the number of clusters by one. Agglomerative clustering starts with all objects in their own cluster and at each stage combines two clusters, so that there is one less cluster. Agglomerative clustering is one of the very few clustering methods that have a deterministic algorithm, and this may explain its popularity. There are many variants on agglomerative clustering, and the manual page for the function `hclust` provides some details. Divisive hierarchical clustering can be performed by using the `diana` function from the **cluster** package.

We first compute the clusterings and then show how to plot them and manipulate the outputs. The `hclust` function returns an instance of the *hclust* class, and `diana` returns an object of class *diana*. These are S3 classes, and hence the objects are lists, with certain named components.

```
> hc1 = hclust(manDist)
> hc2 = hclust(manDist, method="single")
> hc3 = hclust(manDist, method="ward")
> hc4 = diana(manDist)
```

We can plot the resulting dendrograms, and the results are shown in Figure 10.3.

```
> par(mfrow=c(4,1))
> plot(hc1, ann=FALSE)
> title(main="Complete Linkage", cex.main=2)
> plot(hc2, ann=FALSE)
> title(main="Single Linkage", cex.main=2)
> plot(hc3, ann=FALSE)
> title(main="Ward's Method", cex.main=2)
> plot(hc4, ann=FALSE, which.plots=2)
> title(main="Divisive Clustering", cex.main=2)
> par(mfrow=c(1,1))
```

The order in which the leaves are plotted (from left to right) is stored in the slot order. For example, hc1$order is the leaf order in the dendrogram hc1 and hc1$labels[hc1$order] yields the sample labels in the order in which they appear.

Dendrograms can be manipulated using the cutree function. You can specify the number of clusters via the parameter k and the function will cut the dendrogram at the appropriate height and return the elements of the clusters. Alternatively, you can directly specify the height at which to cut via the parameter h.

Exercise 10.3
Cut each of the different clusterings into three clusters. Compare the outputs using, for example, the table *function.*

Although the dendrogram has been widely used to represent distances between objects, it should not be considered a visualization method. Dendrograms do not necessarily expose structure that exists in the data. In many cases they impose a preconceived structure (a tree) on the data, and when that is the case it is dangerous to interpret the observed structure.

Hierarchical clustering creates a new set of between-object distances, corresponding to the path lengths between the leaves of the dendrogram. It is interesting to ask whether these new distances reflect the distances that were used as inputs to the hierarchical clustering algorithm. The cophenetic correlation (e.g., Sneath and Sokal (1973, p. 278)), implemented in the function cophenetic, can be used to measure the association between these two distance measures.

In the code below we show how to compute the cophenetic correlation for complete linkage hierarchical clustering.

```
> cph1 = cophenetic(hc1)
> cor1 = cor(manDist, cph1)
> cor1
[1] 0.524
> plot(manDist, cph1, pch="|", col="blue")
```

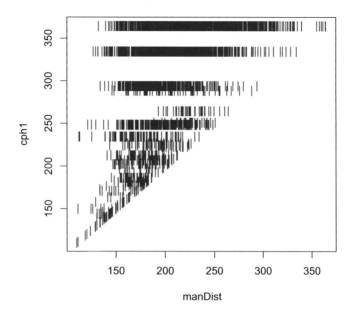

Figure 10.4. A scatterplot of actual distances (on the x-axis) versus the cophenetic distances (on the y-axis) for the hierarchical clustering hc1.

The result is shown in Figure 10.4. The bands in the y direction are due to the discrete nature of the between-sample distances based on trees. For tree-based distances all objects in two subtrees are the same distance from each other.

Exercise 10.4
Compute the cophenetic correlation for the other three dendrograms and comment on which, if any of them, seem to have a particularly good or particularly bad fit.

10.5   Partitioning methods

Let us now turn to partitioning methods. Typically, the algorithms require us to specify the number of clusters into which they should partition the data. There is no generally reliable method for choosing this number, although we may use the estimates we obtained in Section 10.3. Partitioning algorithms have a stochastic element: they depend on an essentially arbitrary choice of a starting partition, which they iteratively update to try to find a good solution.

A simple implementation of a partitioning clustering algorithm, k-means clustering, is provided by the function kmeans. The k-means method attempts to partition the samples into k groups such that the sum of squared distances from the samples to the assigned cluster centers is minimized. The implementation allows you to supply either the location of the cluster centers, or the number of clusters using the `centers` parameter. It is often a good idea to try multiple choices of random starting partitions, which can be specified by the `nstart` parameter. The function returns the partition with the best objective function (the smallest sum of squared distances), but that does not mean that there is not a better partition that has not been tested.

In the code below, we call `kmeans` twice; in both cases we request two groups, but we try 5 different random starts with the first call, and 25 with the second.

```
> km2 = kmeans(gvals, centers=2, nstart=5)
> kmx = kmeans(gvals, centers=2, nstart=25)
```

Exercise 10.5
What values are returned by `kmeans`? Do the two calls find the same clusters?

Exercise 10.6
Which one of the categorical phenotypic variables for our expression set best aligns with the output of the k-means clustering algorithm?

10.5.1   PAM

Partitioning around mediods (PAM) is based on the search for k representative objects, or medoids, among the samples. Then k clusters are constructed by assigning each observation to the nearest medoid with a goal of finding k representative objects that minimize the sum of the dissimilarities of the observations to their closest representative object. This method is implemented by the `pam` function, from the **cluster** package. It is much more flexible than the `kmeans` function in that one can specify different distance metrics to use or supply a distance matrix to use, rather than a data matrix.

```
> pam2 = pam(manDist, k=2, diss=TRUE)
> pam3 = pam(manDist, k=3, diss=TRUE)
```

We can compare the two clusterings, but need to do a little checking to ensure that the orderings are the same.

```
> all(names(km2$cluster) == names(pam2$clustering))
[1] TRUE
> pam2km = table(km2$cluster, pam2$clustering)
> pam2km

     1  2
  1 62  0
  2  3 14
```

Exercise 10.7
*How many items are classified in the same way by the two methods
(k-means and PAM)? How many are classified differently? Can you determine which ones they are? And, then using the cluster centers, as reported
by the different methods, which of the two methods is better? How do the
cluster centers for the two methods compare?*

Exercise 10.8
Repeat Exercise 10.6 for PAM clustering into three groups.

10.6 Self-organizing maps

self-organizing maps (SOMs)[Self-organizing maps (SOMs)] were proposed
by Kohonen (1995) as a simple method for allowing data to be sorted
into groups. The basic idea is to lay out the data on a grid, and to then
iteratively move observations (and the centers of the groups) around on
that grid, slowly decreasing the amount that centers are moved, and slowly
decreasing the number of points considered in the neighborhood of a grid
point. For our examples we use a four-by-four grid, so that there are at
most 16 groups.

We examine two implementations, one in the **kohonen** package, and the
SOM function in the package **class**. The second of these is described in more
detail in Venables and Ripley (2002). We do note that there are others, such
as that provided by the **som** package and readers might want to consider
that version as well. Unfortunately the default values, calling sequences and
return values for these different implementations tend to vary and so you
as a user will need to use some caution in comparing them.

First we demonstrate the use of SOMs using the **kohonen** package. We
fit three different models: the first uses the default values, and the next
two calls change some of these. We set the seed for the random number
generator to ensure that readers get the same answers we do.

```
> set.seed(123)
> s1 = som(gvals, grid=somgrid(4,4))
> names(s1)
 [1] "data"         "grid"          "codes"
 [4] "changes"      "alpha"         "radius"
 [7] "toroidal"     "unit.classif" "distances"
[10] "method"
> s2 = som(gvals, grid=somgrid(4,4), alpha=c(1,0.1),
        rlen=1000)
> s3 = som(gvals, grid=somgrid(4,4, topo="hexagonal"),
        alpha=c(1,0.1), rlen=1000)
> whGP = table(s3$unit.classif)
> whGP

 1  2  3  4  5  6  7  8  9 10 11 12 13 14 15 16
 1  9  2  1  1  1  1  1  7 10  1 19  1  1 15  8
```

The output is an instance of the `kohonen` class. And the last call (to `table`) in the code above, tells us which sample is assigned to which of the 16 possible groups. There are two groups with more than ten observations in them, and nine with only one. The groups with only one are problematic, and although they may represent clusters, it is not clear that they do.

Exercise 10.9
Read the man page for the `kohonen` class. What are the components of this class? How do the results of using the default values compare to the other two?

Another important aspect of understanding the data would be to consider the samples in the different groups and to visualize them.

Exercise 10.10
We choose two of the larger clusters in the output of the first clustering. Create a heatmap comparing those in cluster 13 to those in cluster 14.

Next we consider the `SOM` from the **class** package. This function returns the grid that the map was laid out on, as well as a matrix of representatives; one then uses the `knn1` function to match a sample to its nearest representative. We begin by setting the seed for the random number generator to ensure that readers get the same output as we do.

```
> set.seed(777)
> s4 = SOM(gvals, grid=somgrid(4,4, topo="hexagonal"))
> SOMgp = knn1(s4$code, gvals, 1:nrow(s4$code))
> table(SOMgp)
```

```
SOMgp
  1  2  3  4  5  6  7  8  9 10 11 12 13 14 15 16
  0  1  9  5  1  1  2  0  2 11  1 17  1  3  6 19
```

Now we can see that there are some groups that have no values in them, whereas others tend to have between 10 and 15. To further refine the clusters, down to just a few, we might next ask whether any of the cluster centroids are close to each other, suggesting that merging of the clusters might be worthwhile. We compute the distance matrix comparing cluster centers next, and from that computation we see that clusters (1, 2, 5, 6) can be collapsed, as can (3, 10), (4,15), (7,9), (8, 11, 13, 14). We make this observation based on the zero entries in the distance matrix, cD, computed below.

```
> cD = dist(s4$code)
> cD
           1       2       3       4       5       6       7       8
2   0.000
3   0.857 0.857
4   1.573 1.573 1.813
5   0.000 0.000 0.857 1.573
6   0.000 0.000 0.857 1.573 0.000
7   0.839 0.839 1.132 1.571 0.839 0.839
8   1.182 1.182 1.558 1.796 1.182 1.182 1.219
9   0.839 0.839 1.132 1.571 0.839 0.839 0.000 1.219
10  0.857 0.857 0.000 1.813 0.857 0.857 1.132 1.558
11  1.182 1.182 1.558 1.796 1.182 1.182 1.219 0.000
12  2.669 2.669 3.132 3.046 2.669 2.669 2.888 2.565
13  1.182 1.182 1.558 1.796 1.182 1.182 1.219 0.000
14  1.182 1.182 1.558 1.796 1.182 1.182 1.219 0.000
15  1.573 1.573 1.813 0.000 1.573 1.573 1.571 1.796
16  2.176 2.176 2.445 2.648 2.176 2.176 2.167 2.290
           9      10      11      12      13      14      15
2
3
4
5
6
7
8
9
10  1.132
11  1.219 1.558
12  2.888 3.132 2.565
13  1.219 1.558 0.000 2.565
```

```
14 1.219 1.558 0.000 2.565 0.000
15 1.571 1.813 1.796 3.046 1.796 1.796
16 2.167 2.445 2.290 3.788 2.290 2.290 2.648
```

So, we can then remove the redundant codes and remap the data into clusters using the knn1 function as above.

```
> newCodes = s4$code[-c(2,5,6,10, 15, 9, 11, 13, 14),]
> SOMgp2 = knn1(newCodes, gvals, 1:nrow(newCodes))
> names(SOMgp2) = row.names(gvals)
> table(SOMgp2)
SOMgp2
 1  2  3  4  5  6  7
 3 20 11  4  5 17 19
> cD2 = dist(newCodes)
> cmdSOM = cmdscale(cD2)
```

As we see there are now four reasonably large groups, and three smaller ones.

Exercise 10.11
Compare this clustering with k-means output with k set to 4. What happens if you remove the arrays that correspond to the small clusters and redo the k-means analysis?

10.7   Hopach

The **hopach** package (Pollard and van der Laan, 2005) uses a hybrid approach to clustering. It makes use of both a hierarchical approach as well as a partitioning method. In Section 10.3 we introduced two functions from this package for assessing how many clusters are in the data. In this section we use hopach to cluster our samples.

```
>   samp.hobj = hopach(gvals, dmat = manDist)
>   samp.hobj$clust$k
[1] 3
```

This suggests that there are three clusters. We should first see what sizes they are.

```
> samp.hobj$clust$sizes
[1] 24 28 27
```

We next consider hopach clustering for the genes. In this case, we follow the advice in Pollard and van der Laan (2005) and use the `cosangle` distance between genes.

```
> gene.dist = distancematrix(t(gvals), d = "cosangle")
> gene.hobj = hopach(t(gvals), dmat = gene.dist)
> gene.hobj$clust$k
[1] 40
```

So we see that hopach is suggesting that there are 40 clusters of genes and their sizes are shown in the output below.

```
> gene.hobj$clust$sizes
 [1]  1  1  1  2  1  1  3  1  1  1  1  3  3  1  1  1 11
[18] 57 53 28  1  1  2  1  4  1  1  1  4  1  1  1  1  3
[35]  1  5  1  1  1 45
```

Next one can try to identify important, possibly functional relationships for the genes in the different clusters. A fairly straightforward process would be to use the **GOstats** package to perform an analysis on these genes.

10.8   Silhouette plots

Silhouette plots can be produced using the `silhouette` function in the **cluster** package. It can be defined for virtually any clustering algorithm, and provides a nice way to visualize the output.

The silhouette for a given clustering, C, is calculated as follows. For each item j, calculate the average dissimilarity $\bar{d}_{jl}$ of item j with other genes in the cluster C_l, for all l. Thus, if there are L clusters, we would compute L values for each item. If item j is assigned to cluster l^* then let $a_j = \bar{d}_{jl^*}$, and let $b_j = \min_{l \neq l^*} \bar{d}_{jl}$. The silhouette of item j is defined by the formula: $S_j = (b_j - a_j)/\max(a_j, b_j)$. Heuristically, the silhouette measures how similar an object is to the other objects in its own cluster versus those in some other cluster. Values for S_j range from 1 to -1, with values close to 1 indicating that the item is well clustered (is similar to the other objects in its group) and values near -1 indicating it is poorly clustered, and that assignment to some other group would probably improve the overall results.

We revisit the PAM clusterings, because there are plotting methods for them. These are shown in Figure 10.5.

```
> silpam2 = silhouette(pam2)
> plot(silpam2, main="")
```

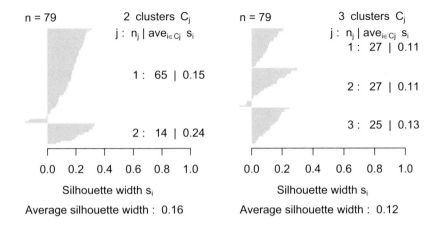

Figure 10.5. Silhouette plots for the PAM clustering of the ALL data: left: two clusters; right: three clusters.

```
> silpam3 = silhouette(pam3)
> plot(silpam3, main="")
```

We can see that there are several samples which have negative silhouette values, and fairly natural questions include "Which samples are these?" and "To what cluster are they closer?" This can be easily determined from the output of the `silhouette` function.

```
> silpam3[silpam3[,"sil_width"] < 0,]
      cluster neighbor sil_width
12026       2        1 -3.97e-05
27003       2        3 -3.21e-02
43004       2        3 -3.30e-02
43001       2        1 -4.15e-02
28031       2        3 -7.80e-02
```

Exercise 10.12
How many samples have negative silhouette widths for the `pam2` *clustering?*

Exercise 10.13
For one of the hierarchical clustering algorithms, divide the data into four clusters and produce a silhouette plot for those four clusters. You will need to read the manual page for the `silhouette` *function to see how to provide the necessary input data.*

10.9 Exploring transformations

Cluster discovery can be aided by the use of variable transformations. We have mentioned multidimensional scaling above in connection with distance assessment. The principal components transformation of a data matrix re-expresses the features using linear combinations of the original variables. The first principal component is the linear combination chosen to possess maximal variance, the second is the linear combination orthogonal to the first possessing maximal variance among all orthogonal combinations, and further principal components are defined (up to p for a rank p matrix) in like fashion. Principal components are readily computed using the singular value decomposition (see the R function `svd`) of the data matrix, and the `prcomp` function will compute them directly. We illustrate the process using the following filtering of the ALL data to 50 genes.

```
> rtt = rowttests(ALLfilt_bcrneg, "mol.biol")
> ordtt = order(rtt$p.value)
> esTT = ALLfilt_bcrneg[ordtt[1:50],]
```

With the raw variables, a five-gene pairwise display is easy to make; we color it with class labels even though we are describing tasks for unsupervised learning (Figure 10.6).

```
> pairs(t(exprs(esTT)[1:5,]),
      col=ifelse(esTT$mol.biol=="NEG", "green", "blue"))
```

Here is how we compute the principal components. We transpose the expression matrix so that gene expression levels are regarded as features of sample objects. In this unsupervised re-expression of the data, clusters corresponding to the different phenotypes are more readily distinguished than they are in the pairwise scatterplot of raw gene expression values (Figure 10.7).

```
> pc = prcomp(t(exprs(esTT)))
```

```
> pairs(pc$x[,1:5], col=ifelse(esTT$mol.biol=="NEG",
      "green", "blue"))
```

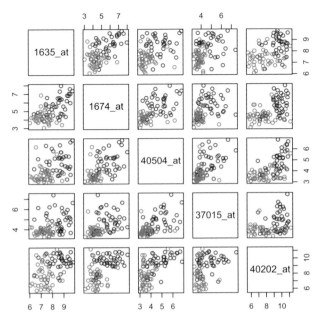

Figure 10.6. A pairs plot for the first five genes in the filtered ALL data.

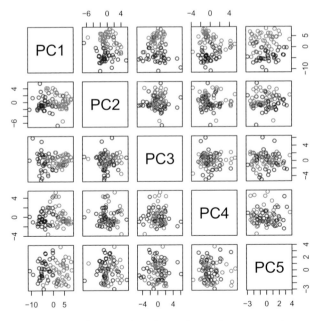

Figure 10.7. A pairs plot for the first five principal components computed from the filtered ALL data.

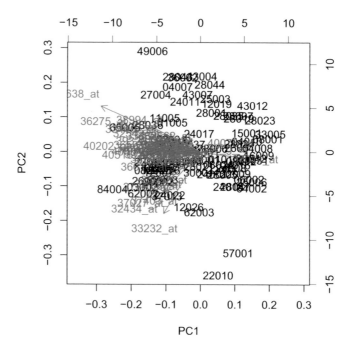

Figure 10.8. A biplot for the first two principal components computed from the filtered ALL data.

Biplots enhance the pairwise principal components display by providing information on directions in which the original variables are transformed to create principal components (Figure 10.8).

```
> biplot(pc)
```

Exercise 10.14
Certain probe set names are prominent in the biplot. Using two-sample tests, explain their roles in discriminating the two phenotypes.

Exercise 10.15
Create a less stringent filtering of the ALL data and generate the associated pairs and biplot displays.

10.10   Remarks

We have given a rudimentary view of the tools available in R for unsupervised machine learning. Most of the ones we have discussed have substantially more capabilities than we have considered and there are many, that are worthwhile that we have not been able to present.

Furthermore, it seems that there is still a great deal of research that can be done in this area. Current topics that need to be addressed are the detection of outlying items and the development of tools that can use additional genomic information in developing and devising the clustering (we gave a simple example, because we concentrated on features obtained from specific GO categories).

11

Using Graphs for Interactome Data

T. Chiang, S. Falcon, F. Hahne, and W. Huber

Abstract

Many data types and many models of biological systems are best described in terms of graphs. Protein–protein interaction data are a prominent example. In this chapter, we explore a curated dataset of protein interactions and perform a statistical analysis of the relationship between protein interaction and coexpression. We also show how to access large-scale protein–protein interaction datasets from the *IntAct* repository at the EBI.

11.1   Introduction

There are three main packages in Bioconductor that handle graphs: the package **graph** for data structures, **RBGL** for algorithms, and **Rgraphviz** for graph layout. A more detailed description of their capabilities can be found in Carey et al. (2005). In the first part of this chapter, we consider a protein interaction dataset, perform some initial data exploration, and show that there is a statistically significant association between the fact that two proteins physically interact and that they appear coexpressed based on the data from a microarray experiment. The analysis follows section 22.2 of Gentleman et al. (2005a), which in turn is based on Ge et al. (2001) and Balasubramanian et al. (2004).

We create two graphs, one whose edges indicate that two genes are in the same co-expression cluster and the other where an edge indicates that their protein products physically interact. If there is an association between co-expression and protein interaction, then we anticipate that when there is an edge between two genes in one graph, there will also be an edge between them in the other graph. We can test this hypothesis by counting how many edges the two graphs have in common and comparing this number to how

F. Hahne et al., *Bioconductor Case Studies*, DOI: 10.1007/978-0-387-77240-0_11,
© Springer Science+Business Media, LLC 2008

much overlap we would expect "by chance". More precisely, we generate a
reference distribution by randomly permuting the node labels on either one
of the two graphs and for each permutation count the common edges. The
comparison of the observed number of common edges with the permutation
distribution gives us an indication of how significant the overlap is.

First, we load the packages that we need for this chapter.

```
> library("Biobase")
> library("graph")
> library("Rgraphviz")
> library("RColorBrewer")
> library("RBGL")
> library("yeastExpData")
> library("Rintact")
```

The **yeastExpData** package contains two curated datasets: `ccyclered`
contains co-expression clusters obtained from a microarray experiment
measuring gene expression during the yeast cell-cycle; `litG` contains
protein–protein interactions (PPI) extracted from published papers.

```
> data("ccyclered")
> data("litG")
```

Exercise 11.1

a. *What type of object is* `litG`? *How do you find out more about this
class?*

b. *Use the* `nodes` *method to extract the first five nodes of* `litG`.

c. *Explore the* `ccyclered` *data to determine what type of R object it is
and what kind of data it contains.*

11.2   Exploring the protein interaction graph

A graph consists of one or more connected components. You can find them
using the `connectedComp` function from the **RBGL** package.

```
> cc = connectedComp(litG)
> length(cc)
[1] 2642
> cclens = sapply(cc, length)
> table(cclens)
```

```
cclens
   1     2     3     4     5     6     7     8    12    13    36    88
2587    29    10     7     1     1     2     1     1     1     1     1
```

Exercise 11.2

  a. What are the elements of cc?

  b. How many connected components are there? What is the size of the
     largest connected component? How many singletons are there?

We can use the subGraph function to create two new graphs sg1 and
sg2 that represent the largest and second largest connected components of
litG.

```
> ord = order(cclens, decreasing=TRUE)
> sg1 = subGraph(cc[[ord[1]]], litG)
> sg2 = subGraph(cc[[ord[2]]], litG)
```

Now we plot sg1 and sg2 using **Rgraphviz**. The layout step is done
using function layoutGraph and function renderGraph is responsible for the
subsequent plotting. There are many options for the color and type of the
nodes and edges. If you are interested in the details or in producing more
complex graphics, please refer to Section 12 and the documentation of the
Rgraphviz package.

```
> lsg1 = layoutGraph(sg1, layoutType="neato")
> lsg2 = layoutGraph(sg2, layoutType="neato")
```

```
> renderGraph(lsg1)
```

```
> renderGraph(lsg2)
```

The result is shown in Figure 11.1.

Exercise 11.3
Lay out the graphs using the dot and twopi layout engines.

Next, let us compute the shortest path between a pair of nodes in the
largest component. This can be a useful computation in some applications.
In a protein–protein interaction graph it has no apparent biological interpretation, and we do it only to illustrate this piece of functionality of the
RBGL package.

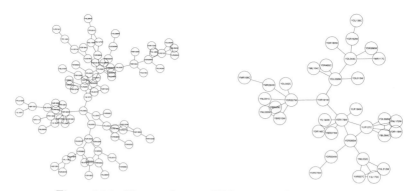

Figure 11.1. The two largest PPI connected components.

```
> sps = sp.between(sg1, "YHR129C", "YOL039W")
> pth = sps[[1]]$path_detail
> pth
 [1] "YHR129C" "YPL174C" "YOR098C" "YOR160W" "YAL005C"
 [6] "YAL040C" "YJL157C" "YBR200W" "YFL039C" "YDR382W"
[11] "YLR340W" "YOL039W"
```

Exercise 11.4

What sort of object is sps*? What does the manual page say about it? Can you plot the graph and identify that this indeed is the shortest path? (You could color these nodes differently from the rest as shown in Figure 11.2.)*

The *diameter* of a graph is defined as the length of the longest shortest path between any two nodes. To compute this we use the function johnson.all.pairs.sp.

```
> allp = johnson.all.pairs.sp(sg1)
```

Exercise 11.5

What type of object is allp*? What data does it contain? What is the diameter of* sg1*? Is the longest shortest path unique in this graph?*

11.3   The co-expression graph

Let us now have a look at the co-expression clusters. Cho et al. (1998) presented the k-means clustering of 2885 *Saccharomyces cerevisiae* genes into 30 clusters with measurements taken over two cell cycles. Their clustering is stored in the dataframe ccyclered. Each row corresponds to a gene, and its cluster membership is indicated by integer numbers from 1 to 30 in the

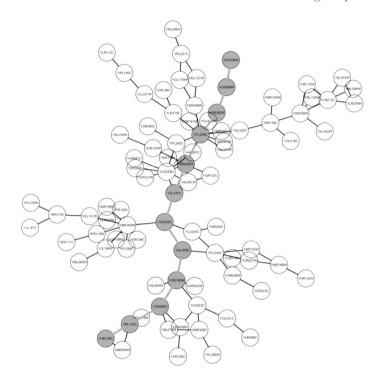

Figure 11.2. The largest connected component of `litG` and the shortest path between nodes YHR129C and YOL039W.

column `Cluster`. Our first step is to create a cluster graph in which edges are between all genes that are in the same cluster. There is a specialized graph class *clusterGraph* that can be used to describe such graphs. We need to compute the set of genes in each cluster, and we do that by building a list in which each entry represents a cluster and consists of a character vector with the names of the genes in the cluster.

Exercise 11.6
Use the `split` *function and the* `Y.name` *and* `Cluster` *columns of the* `ccyclered` *data frame to create a named list* `clusts` *whose elements correspond to the clusters, such that each list element is a vector of gene names in that cluster.*

Next we use the `clusts` list from the previous exercise to create a *clusterGraph* instance using `new`:

```
> cg = new("clusterGraph", clusters = clusts)
```

Exercise 11.7
How many connected components does the cluster graph `cg` *have?*

11.4   Testing the association between physical interaction and coexpression

It is now easy to determine how many pairs of genes have both a protein–protein interaction and are found in the same expression cluster. To compute this, find the intersection of the cluster graph and the literature graph using `intersection`:

```
> commonG = intersection(cg, litG)
```

Exercise 11.8
How many edges are common to the two graphs (cg and `litG`)?

Now we try to determine whether the number of common edges is statistically interesting. We do this by generating a null distribution via permutation of node labels on the observed graph. Here is a function that can be used to generate values from the desired null distribution. Unfortunately, running this function with the current implementation is very slow.

```
> nodePerm = function (g1, g2, B=1000) {
      n1 = nodes(g1)
      sapply(1:B, function(i) {
          nodes(g1) = sample(n1)
          numEdges(intersection(g1, g2))
      })
  }
```

Exercise 11.9
Describe what the `nodePerm` function is doing to make sure you understand how it works.

Because the `nodePerm` function is slow, we've computed 500 iterations ahead of time. Load the precomputed result as follows:

```
> data("nPdist")
> summary(nPdist)
```

Exercise 11.10
Plot the `nPdist` data and decide if the number of edges in common between `litG` and `cg` is statistically interesting.

11.5   Some harder problems

In this section we present some problems that are more open-ended. They are not formally part of this chapter, but are here for those who are particularly interested in these sorts of applications. To answer the last two questions you need to obtain a Bioconductor package called **ScISI**.

- Which of the expression clusters have intersections with the literature clusters?

- Are there expression clusters that have many literature cluster edges going between them? (This could suggest that the expression clustering was too fine, or that the co-citations of the genes in the literature are not cell-cycle related.)

- Do the genes involved in known cell-cycle regulated protein complexes tend to cluster together in both graphs?

- Is the expression behavior of genes that are involved in multiple protein complexes different from that of genes that are known to be involved in only one complex?

11.6   Reading PSI-25 XML files from *IntAct* with the **Rintact** package

11.6.1   Introduction

Rintact is an R package mainly used to parse the PSI-25 files generated by the *IntAct* data repository. *IntAct* collects, curates, and stores thousands of protein interactions. Currently, there are three main functions within **Rintact**:

1. `psi25interaction`

2. `psi25complex`

3. `intactXML2Graph`

The first function, `psi25interaction`, takes either a PSI-MI 2.5 XML file from *IntAct* or an URL containing the Web address of where such an XML file can be obtained. The XML file must contain binary protein–protein interaction data. Examples for such data are direct physical interactions, complex co-membership, and synthetic genetic interactions. The second function, `psi25complex`, also takes a PSI-MI 2.5 XML file or URL as an input parameter, but this file must contain protein complex membership information. The third function, `intactXML2Graph`, can either take an interaction or complex XML file and returns a *graph* data structure upon which analysis can be made. In principle, these three functions can take any XML file that adheres to the PSI-MI 2.5 standards. We have constructed these functions, however, to work primarily with the *IntAct* PSI-MI 2.5 XML files, as there are subtle implementation differences between repositories such as *IntAct* and *DIP*, both of which use the PSI-MI 2.5 standards. In this vignette, we demonstrate the use of these functions on sample PSI-MI 2.5 datasets for both protein interactions as well as the manually curated protein complexes.

11.6.2   Loading R Packages

We begin by loading the various R packages which we use. Our primary focus is with the **Rintact** package, but we also examine and exploit statistical methods found in various Bioconductor packages for the analysis of the interaction data obtain from *IntAct*.

```
> library("Rintact")
> library("ppiStats")
> library("apComplex")
> library("xtable")
```

11.6.3   Obtaining the interaction information

We first demonstrate the use of the function psi25interaction. We can either download the *IntAct* PSI-MI 2.5 XML file to a local file system, or we can simply use the URL (of where the file can be obtained) as the input parameter. Here, we use a file that is part of the examples which come with the **Rintact** package.

```
> fn = system.file("PSI25XML", "interactionSample.xml",
      package="Rintact")
> eg = psi25interaction(fn)
> class(eg)
[1] "interactionEntry"
attr(,"package")
[1] "Rintact"
```

We see that the output of psi25interaction is an instance of the class *interactionEntry*. This class has five slots:

```
> slotNames(eg)
[1] "organismName" "taxId"        "releaseDate"
[4] "interactors"  "interactions"
```

Three of them contain simple character vectors:

```
> organismName(eg)
[1] "Homo sapiens"
[2] "Human adenovirus E"
[3] "Human papillomavirus type 1a"
> taxId(eg)
[1] "9606"    "130308" "10583"
> releaseDate(eg)
[1] "2007-04-27"
```

organismName records all the organisms for which interactions were assayed. For each organism, we have also included its taxonomy identification code. Because *IntAct* does not currently version its weekly release, we have added the *releaseDate* as a time stamp to act as a surrogate version number.

Let us investigate the structure of the *interactions* slot. It is a list that holds all the binary interactions given in the XML file, along with information about each particular interaction. Each element of the list is an instance of the class *intactInteraction*. This class has nine slots:

```
> length(interactions(eg))
[1] 5
> class(interactions(eg)[[1]])
[1] "intactInteraction"
attr(,"package")
[1] "Rintact"
> interactions(eg)[[1]]
interaction ( EBI-987168 ):
  --------------------------------
  [ interaction type ]: pull down
  [ experiment ]: pubmed  16249186 , intact  EBI-965562
  [ confidence value ]:  NA
  [ bait ]:  EBI-491274
  [ prey ]:  EBI-448924
  [ neutral component ]:  NA
  [ inhibitor ]:  EBI-987160
> slotNames(interactions(eg)[[1]])
[1] "intact"              "interactionType"  "expPubMed"
[4] "expIntAct"           "confidenceValue"  "bait"
[7] "prey"                "inhibitor"        "neutralComponent"
```

The various slots contain information that is relevant for each individual interaction. `interactionType` details what manner of interaction was found between the bait protein and the prey protein, which are specified in the `bait` and `prey` slots. Another attribute is the experimental confidence value given in the `confidenceValue` slot. This confidence value is reported by the experimenters; it does not report scores derived by third parties.

We can extract the names of the bait and prey proteins for all of the interactions in the `eg` dataset:

```
> egbait = sapply(interactions(eg), bait)
> egprey = sapply(interactions(eg), prey)
```

We now have two character vectors, `egbait` and `egprey`, that are aligned with each other: the ith protein in `egprey` is found by the ith protein in `egbait`.

```
> egbait
[1] "EBI-491274" "EBI-491274" "EBI-963841" "EBI-491274"
[5] "EBI-963841"
> egprey
[1] "EBI-448924" "EBI-448924" "EBI-491274" "EBI-448924"
[5] "EBI-765551"
```

The *IntAct* accession codes are useful as unique and uniform identifiers in the *IntAct* repository, but we usually want to translate them to other identifier schemes such as the HUGO gene name or Ensembl gene identifier. The PSI-MI 2.5 XML files from *IntAct* contain a lookup table for this purpose. This lookup table is stored in the *interactors* slot of the *interactionEntry* object `eg`, in the form of a character matrix. Its rows are indexed by the *IntAct* accession numbers of the molecules in the data structure, and it has seven columns.

```
> interactors(eg)
            uniprotId geneName
EBI-491274  "P06400"  "RB1"
EBI-987160  "Q6H1D8"  "E1A"
EBI-448924  "Q01094"  "E2F1"
EBI-963841  "P06465"  "E7"
EBI-765551  "O00716"  "E2F3"
            fullName                            locusName
EBI-491274  "Retinoblastoma-associated protein" NA
EBI-987160  "E1A"                               NA
EBI-448924  "Transcription factor E2F1"         NA
EBI-963841  "Protein E7"                        NA
EBI-765551  "Transcription factor E2F3"         NA
            orfName organismName           taxId
EBI-491274  NA      "Homo sapiens"         "9606"
EBI-987160  NA      "Human adenovirus E"   "130308"
EBI-448924  NA      "Homo sapiens"         "9606"
```

```
EBI-963841 NA       "Human papillomavirus type 1a" "10583"
EBI-765551 NA       "Homo sapiens"                 "9606"
```

The *IntAct* accession codes can be translated into any of the associated identifier schemes. Two further properties are given for each molecule: the organism in which the molecule is native and the corresponding taxonomy ID. Most of the interactions found in *IntAct* will be protein–protein interactions; other types of interactions, however, are also stored such as small molecule-to-protein interactions as well as gene–gene interactions. As a result, there will be times when a molecule cannot be mapped to a locus name or an ORF, and so on. We also remark that interactions have been tested between proteins of different organisms (e.g., a human protein against a viral one). Thus the organism attribute is vital to keep such interactions in the proper context.

Using the lookup table is quick and efficient because of the subsetting functionality of R. For instance, say we would like to translate the following *IntAct* accession codes

```
> bts = egbait[3:4]
```

into gene names:

```
> interactors(eg)[bts, c("geneName","fullName")]
           geneName fullName
EBI-963841 "E7"      "Protein E7"
EBI-491274 "RB1"     "Retinoblastoma-associated protein"
```

11.6.4   Obtaining protein complex composition information

Now we demonstrate the parser function psi25complex.

The parameters of psi25complex are identical to those of psi25interaction, althpugh its output only contains three slots:

```
> fn2 = system.file("PSI25XML/complexSample.xml",
     package="Rintact")
> comps = psi25complex(fn2)
> slotNames(comps)
[1] "releaseDate" "interactors" "complexes"
```

Again releaseDate serves as a surrogate version number, and the interactors holds a lookup table that can be used to translate the *IntAct* accession codes. The complexes slot is a list of *intactComplex* objects. Each list entry is an instance of the class *intactComplex*, which itself has seven slots.

```
> length(complexes(comps))
[1] 174
> class(complexes(comps)[[1]])
[1] "intactComplex"
attr(,"package")
[1] "Rintact"
> slotNames(complexes(comps)[[1]])
[1] "intactId"     "shortLabel"   "fullName"
[4] "organismName" "taxId"        "members"
[7] "attributes"
```

These slots describe the multiprotein complex. The three most important ones are fullName, attributes, and members slots. The fullName slot gives the exact name of the multiprotein complex and the attributes slot gives a short description as to the known functionality of the complex. The interactors slot gives the members of the complex and their multiplicity.

```
> complexes(comps)[[1]]
complex ( EBI-706546 )
----------------------------------
 [ short label ]:  bcl2_bcl2_human
 [ full name ]:  BCL-2 homodimer
 [ organism ]:  Homo sapiens
 [ taxonomy ID ]:  9606
 [ attributes ]:
curated-complex: Role of homodimer is unclear, may act as rese
  rvoir of protein for heterodimer formation.
complex-synonym: Bcl2 homodimer; Bcl-2:Bcl-2; Bcl2:Bcl2;
kd: 0.0
 [ members ]:
   intActId multiplicity
4 EBI-77694               2
```

11.6.5 Creating graph objects with Rintact

Now we investigate the function intactXML2Graph. Whereas the functions psi25interaction and psi25complex allow us to parse the PSI-MI 2.5 XML files and to obtain all the information available for protein interaction and complex data, intactXML2Graph takes either an XML file, or a URL to such a file, and converts this into an R *graph* object. Much of the information obtained from either psi25interaction or psi25complex will be muted as it is not necessary in the graph representation.

```
> s1 = system.file("PSI25XML", "human_stelzl-2005-1_01.xml",
      package="Rintact")
> s2 = system.file("PSI25XML", "human_stelzl-2005-1_02.xml",
      package="Rintact")
> stelzl = intactXML2Graph(c(s1,s2), type="interaction",
      directed=TRUE)
> class(stelzl)
> stelzl1 = removeSelfLoops(stelzl)
```

Often, *IntAct* will take one dataset and split it into multiple XML files. The intactXML2Graph function can combine all of the PSI-MI 2.5 files from a single experiment and construct the graph from the amalgamation of these files. The output from intactXML2Graph is an *intactGraph* object. If the XML file contained protein complex membership data, the output would be an instance of the *intactHyperGraph* object.

We can now investigate the human protein–protein interaction graph by rendering a subset of the bait-to-prey relationships. First we calculate the degrees of the directed bait to prey graph stelzl1 after having removed self-loops. The deg object is a list that contains both the in- and out-degree statistic for each protein. We define a subset of the bait which we call the activeBait by imposing that these bait find at least 10 different prey and at most 15 different prey. On the union of the activeBait and their prey, we can generate a bait-to-prey subgraph (stelzlSG) and render its pictorial representation.

```
> deg = degree(stelzl1)
> activeBait = names(which(deg$outDegree > 10 &
      deg$outDegree<15))
> proteins = union(activeBait, unlist(adj(stelzl1,
      activeBait)))
> stelzlSG = subGraph(proteins, stelzl1)
```

To differentiate between a bait protein and a prey protein in stelzlSG, we color each vertex representing a protein from the activeBait subset yellow and each of the prey is colored blue.

```
> graph.par(list(nodes=list(fill="steelblue", label="")))
> baitCol = rep("yellow", length(activeBait))
> names(baitCol) = activeBait
> nodeRenderInfo(stelzlSG) <- list(fill=baitCol)
```

It is of interest to note that the graph stelzlSG obtained on the adjacency of the activeBait proteins forms a single connected component (Figure 11.3).

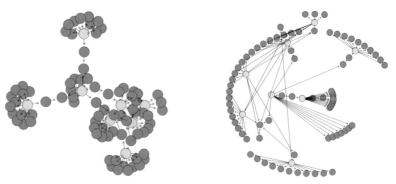

Figure 11.3. These two (equivalent) graphs shows the bait which found between 10 and 15 different prey proteins. Each of the bait proteins has been colored yellow and the prey proteins colored blue. The graph in the left panel has been rendered in the *neato* layout whereas the graph in the right panel has been rendered in the *twopi* layout.

```
> stelzlSG <- layoutGraph(stelzlSG, layoutType="neato")
> renderGraph(stelzlSG)
> stelzlSG <- layoutGraph(stelzlSG, layoutType="twopi")
> renderGraph(stelzlSG)
```

12

Graph Layout

F. Hahne, W. Huber, and R. Gentleman

Abstract

In this chapter we demonstrate how to lay out and render graphs using tools from the **Rgraphviz** and **graph** packages.

12.1   Introduction

Laying out and rendering a graph is important to aid in understanding the relationships within it. However, these tasks are not trivial. The spatial position of each node in the layout of a graph depends on the location of every other node in the plot and on the edges between them. Often we aim to minimize the number of edges that cross. In addition, nodes as well as edges can be represented with different symbols, colors, and the like. There are two distinct processes: *layout*, which places nodes and edges in a (usually two-dimensional) space, and *rendering*, which is the actual drawing of the graph on a graphics device. The first process is typically the more computationally expensive and relies on sophisticated algorithms that arrange the graph's components based on different criteria. The arrangement of the nodes and edges depends on various parameters such as the desired node size, which again may be a function of the size of the node labels. Rendering of a graph is often subject to frequent changes and adaptions, and it makes sense to separate the two processes in the software implementation. It is also important to realize that the process of getting a good layout is iterative, and using default parameter settings seldom yields good plots.

The code available for doing graph layout in Bioconductor is based mainly on the *Graphviz* project (Gansner and North, 1999) and the *Boost graph library* (Siek et al., 2002). However, because the rendering of a graph is separated from the layout, one can use other graph layout algorithms, as long as the requirements of the rendering interface are met.

F. Hahne et al., *Bioconductor Case Studies*, DOI: 10.1007/978-0-387-77240-0_12,
© Springer Science+Business Media, LLC 2008

In the process of laying out a graph layout some amount of information is generated, mostly regarding the locations and dimensions of nodes on a two-dimensional plane and the trajectories of the edges. Bioconductor *graph* objects contain a slot `renderInfo` to hold this information. The typical workflow of a graph layout is to pass a *graph* object to the layout function, which returns another *graph* object containing all the necessary information for subsequent rendering. The process of calling a layout algorithm is encapsulated in the `layoutGraph` function. Calling this function without any further arguments will result in using one of the *Graphviz* layout algorithms via the the the **Rgraphviz** package. There are a number of parameters for fine-tuning of the graph layout and we introduce some of them later in the chapter.

The rendering of a graph relies solely on R's internal plotting capabilities. As for all other plotting functions in R, many parameters controlling the graphical output can be tuned. However, because there are several parts of a graph one might want to modify (e.g., nodes, edges, captions), setting the graphical parameters is slightly more complex than for other plots. We have established a hierarchy to set global defaults, graph-specific parameters, and settings that apply only to individual rendering operations. In the course of this chapter we show the different ways to control the graph rendering. There is much more information about graph rendering in chapters 21 and 22 of Gentleman et al. (2004) .

In order to explore the capabilites for plotting graphs in R we initially use a protein–protein interaction (PIP) dataset containing data that were extracted from biomedical literature. It is part of the **yeastExpData** package and we can load it as a predefined *graph* object:

```
> library("yeastExpData")
> data("litG")
> litG
A graphNEL graph with undirected edges
Number of Nodes = 2885
Number of Edges = 315
```

The whole graph is too big for demonstrating the plotting capabilites of **Rgraphviz**, and we want to focus on something more compact. For this purpose we can choose one of the graph's smaller connected components using the functions `connectedComp` from **RGBL** and `subGraph`. For now let us use a connected subgraph containing 12 nodes:

```
> library("RBGL")
> cc = connectedComp(litG)
> len = sapply(cc, length)
> table(len)
```

```
len
    1     2     3     4     5     6     7     8    12    13    36    88
 2587    29    10     7     1     1     2     1     1     1     1     1
> ord = order(len, decreasing=TRUE)
> g = subGraph(cc[[ord[4]]], litG)
```

Instead of the fourth-largest, you could choose other connected components of different sizes for the examples in this section and compare the results.

12.2   Layout and rendering using **Rgraphviz**

The *Graphviz* library offers five different layout algorithms. Look at the documentation of `agopen` in **Rgraphviz**, or of the *Graphviz* library, for a more detailed description of these algorithms. *dot* is a hierarchical layout algorithm for directed graphs with four main phases: cycles are broken, nodes are assigned to layers, nodes are rearranged in layers to minimize edge crossings, and finally edges are computed as splines. *neato* is a layout algorithm for undirected graphs and is closely related to multidimensional scaling. It creates a virtual physical model and optimizes for low-energy configurations. It was recently augmented with a scalable stress majorization algorithm. *twopi* is a circular layout. Basically, one node is chosen as the center and put at the origin. The remaining nodes are placed on a sequence of concentric circles centered about the origin, each a fixed radial distance from the previous circle. All nodes adjacent to the center node are placed on the first circle; all nodes adjacent to the node on the first circle except the center node are placed on the second circle; and so forth. *fdp* implements the Fruchterman–Reingold heuristic including a multigrid solver that handles larger graphs and clustered undirected graphs. Finally, *circo* is a circular layout suitable for graphs with multiple cyclic structures.

We can lay out our graph without setting any further options using the `layoutGraph` function and immediately plot it with `renderGraph`. The default layout algorithm used is *dot*.

```
> library("Rgraphviz")
> x = layoutGraph(g)
> renderGraph(x)
```

The result is shown in the left panel of Figure 12.1.

12.2.1   Rendering parameters

The layout of a graph is the initial step of graph plotting, but here we start with the setting of parameters in the rendering step. You will soon find out

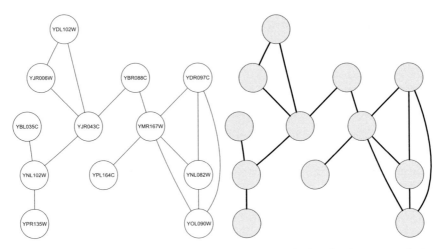

Figure 12.1. A plot of **g** using the **Rgraphviz** defaults for layout and the default global rendering settings (left) and the same graph rendered with user-defined global settings.

that the two processes, although separated in the software implementation, are tightly connected. There is a hierarchy to set rendering parameters for a graph. The levels of this hierarchy are

1. The session: These are the defaults that will be used for a parameter if not set somewhere further down the hierarchy. You can change the session defaults at any time using the function `graph.par`.

2. Rendering operation: Defaults can be set for a single rendering operation, that is, a call to `renderGraph` using its `graph.pars` argument.

3. Individual nodes or edges: Parameters for individual nodes or edges can be set using the `nodeRenderInfo` and `edgeRenderInfo` functions.

We now use our example dataset to further explore these options. The result shown in the left panel of Figure 12.1 is very basic. The default settings produce simple output with both nodes and edges drawn in black and nodes labeled with their names. The node names are often too long for a useful display and the software has to use a tiny font size to make them fit. For now we may decide not to plot node names at all and we can set a global parameter with `graph.par` to do that; the easiest way is to define an empty string. We might also decide to fill all nodes with gray color and to increase the line width of the edges a bit. Note the structure that is used to set the parameters, with list items `nodes`, `edges`, and `graph` (the latter is not used in this instance). This structure allows us to control the rendering of nodes, edges, and graphwide features separately.

```
> graph.par(list(nodes=list(label="", fill="lightgray"),
                  edges=list(lwd=3)))
> renderGraph(x)
```

We can revert `graph.par` changes by setting the respective list item to NULL.

```
> graph.par(list(nodes=list(label=NULL)))
```

Instead of defining global settings with `graph.par` we could also provide a list with the same structure to `renderGraph` through its `graph.pars` argument. Those will only be applied in the respective rendering operation, whereas options set using the function `graph.par` are retained throughout the whole R session.

Exercise 12.1
Try laying out the graph using Graphviz's four other layout algorithms as shown in Figure 12.2. `layoutGraph` *has an argument* `layoutType` *to do that. Also add a title to the plots stating the algorithm that you used. The parameter that controls the title is a graph-specific parameter and it is called* `main`*. Which of the layouts do you prefer for g?*

One might be interested in highlighting certain nodes or in coding categories of nodes or edges by certain colors (Figure 12.3). To this end, options for individual nodes and edges can be set using the `nodeRenderInfo` and `edgeRenderInfo` functions. Both `nodeRenderInfo` and `edgeRenderInfo` are replacement functions that operate directly on the *graph* object. When you change a parameter in the *graph* object this will be carried on across all further rendering and layout operations. The settings made by `edgeRenderInfo` and `nodeRenderInfo` take precedence over all default settings.

The parameters to be set have to be given as named lists, where each list item can contain named vectors for certain options. For example, the following code sets the fill color of nodes YBR088C and YDR097C to red.

```
> nodeRenderInfo(x) = list(fill=c(YBR088C="red",
                           YDR097="red"))
```

The names of the vectors have to match the node or edge names of the graph. Node names are straightforward (the result of calling the function `nodes` on a *graph* object), however edge names are made up of the names of the connected nodes separated by ~, the tilde symbol. An edge between nodes A and B would be named A~B (for a directed graph A~B is the edge fom A to B, and B~A is the edge from B to A). For undirected graphs the two are equivalent. `edgeNames` returns the names of all edges in a graph. The

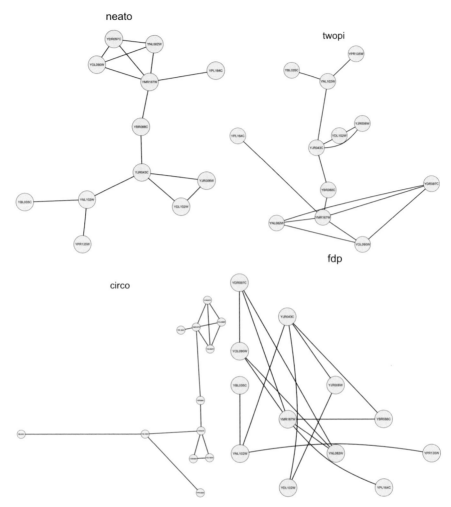

Figure 12.2. These plots show g laid out with **Rgraphviz** using the *neato*, *twopi*, *circo*, and *fdp* options.

following code changes the line type of the edges between nodes YMR167W and YBR088C and nodes YMR167W and YBR088C to "dotted".

```
> edgeRenderInfo(x) = list(lty=c("YBR088C~YMR167W"="dotted",
                               "YDR097C~YMR167W"="dotted"))
```

Exercise 12.2
Change the text color and the border color of some of the nodes. Increase the line width for some of the edges. You can find a complete list of these parameters in the documentation to nodeRenderInfo *and* edgeRenderInfo *or in the vignette of the* **Rgraphviz** *package.*

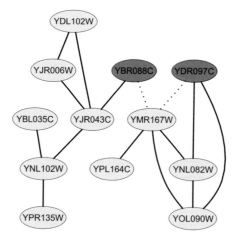

Figure 12.3. g laid out and rendered with user-defined layout and rendering attributes.

12.2.2   Layout parameters

Many of the parameters one may want to change for a graph have an impact on the actual layout. For instance, changing the size of a node may affect the position of neighboring nodes and of the connecting edges. Thus, all parameters that have an influence on the layout have to be given to the layout algorithm. For *Graphviz* there are a multitude of layout parameters, some specific only to certain layout types such as *dot* or *neato*. All these parameters can be specified in the call to layoutGraph. You may want to consult the documentation of the **Rgraphviz** package for more details. Parameters can either be passed as nested named lists, similar to the input to graph.par, or individually for nodes and edges as named lists through the edgeAttrs and nodeAttrs arguments. In order to better fit the node names we can tell *Graphviz* to use ellipses instead of circles and to adjust the sizes of the nodes to accommodate the node labels.

```
> x = layoutGraph(x, attrs=list(node=list(shape="ellipse",
    fixedsize=FALSE)))
> renderGraph(x)
```

Exercise 12.3
What are the different shapes in **Rgraphviz** *that can be used for node drawing? What does the* drawNodes *argument of* renderGraph *do?*

12.3   Directed graphs

In the previous example the direction of an edge was not important. Whenever two proteins were reported to be interacting in a paper they were connected by an edge. In directed graphs the edge from node **A** to node **B** is distinct from the edge from node **B** to node **A**. When plotting, the direction of an edge is usually indicated by little arrows. There may, however, be reciprocated edges and we show in the next section how to deal with them.

   In the remainder of this chapter we use a directed graph as an example to show a typical workflow for producing informative and appealing plots. The data are a graph representation of the integrin-mediated cell adhesion (IMCA) pathway as provided by the KEGG pathway database. In KEGG, pathways are collections of proteins, protein complexes, and processes that are shown to be interacting in the same biological function (Kanehisa and Goto, 2000). Functions comprise cellular processes such as cell division or apoptosis, biochemical reactions, and even certain diseases. Integrin-mediated cell adhesion is one of the processes that keeps cells attached to each other in tissues. A large number of proteins, protein complexes, and subprocesses are involved in the regulation of this pathway. KEGG also offers graphical representations of their pathways and in Figure 12.4 we can see the KEGG image of the IMCA pathway. There are different features used in this plot such as colors, shapes, and positions of nodes that encode a lot of the pathway's structure. The cellular location of the pathway components is resembled by the positioning of the nodes, with the cell membrane components on the left side of the plot whereas on the right side there are the more general cellular processes that feed into the pathway. Note that the actual graph structure tells you only about the connections between nodes, and all additional information is added

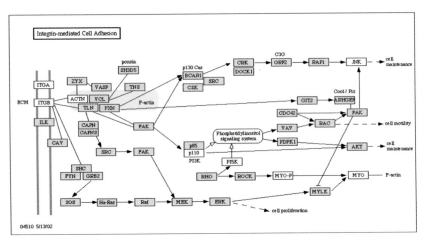

Figure 12.4. The integrin-mediated cell adhesion pathway as rendered by KEGG

through the controlled layout and the rendering. We now want to use the infrastructure in R to produce a similar output. Let us first naively plot our IMCA graph and see how this compares to the KEGG rendering.

```
> data("integrinMediatedCellAdhesion")
> IMCAGraph
A graphNEL graph with directed edges
Number of Nodes = 52
Number of Edges = 91
> IMCAGraph = layoutGraph(IMCAGraph)
> renderGraph(IMCAGraph)
```

The result in Figure 12.5 is not impressive. The node labels are unreadable because the font size is too small, there is no color or grouping, and the layout is not structured to resemble the cellular architecture. As mentioned

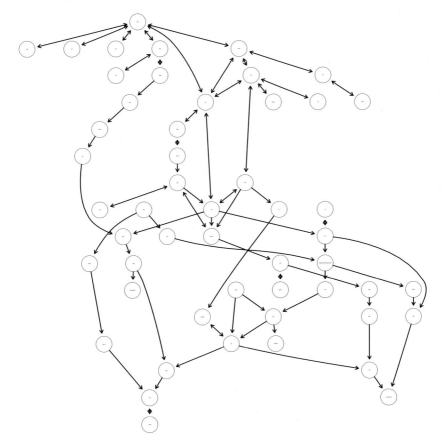

Figure 12.5. The integrin-mediated cell adhesion pathway as rendered by **graphviz** using default settings.

before, producing good graph plots is an art and needs a considerable amount of fine tuning. In R we have the tools to do this tuning and we can iteratively improve the outcome. At each step you should render the graph to check the results of your settings.

Let us first start with the node labels. The default behavior of render-Graph is to compute a common font size for all node labels in a way that they all fit their node. There are some long node names in the graph, and we can determine the length of each node name using the function nchar.

```
> names(labels) = labels = nodes(IMCAGraph)
> nc = nchar(labels)
> table(nc)
nc
 3  4  5  6  7 13 16 18 37
25 13  7  2  1  1  1  1  1
> long = labels[order(nc, decreasing=TRUE)][1:4]
> long
  Phosphatidylinositol signaling system
"Phosphatidylinositol signaling system"
                      cell proliferation
                    "cell proliferation"
                        cell maintenance
                      "cell maintenance"
                           cell motility
                         "cell motility"
```

We need to deal with these four long names specially. One option would be to use an alternative name, maybe an abbreviation. Alternatively, we could include line feeds into the strings in order to force multi-line text. This is what we do. The escape sequence for line feeds in R is \n.

```
> labels[long] = c(paste("Phosphatidyl-\ninositol\n",
      "signaling\nsystem", sep=""), "cell\nproliferation",
      "cell\nmaintenance", "cell\nmotility")
```

Because we want to change a property of individual nodes we have to use nodeRenderInfo for the setting. As shown before, the function matches rendering parameters by the name of the list item and nodes by the names of the items of the individual vectors. The parameter we want to modify is label.

```
> nodeRenderInfo(IMCAGraph) = list(label=labels)
> renderGraph(IMCAGraph)
```

The four labels are now plotted as multi-line strings but this has not changed the layout. Remember that rendering and layout are two distinct processes, and for changes to affect the latter you have to re-run `layout-Graph`. Another layout change we may want to do at this point is to further increase the size of the nodes with long names to give them a little bit more room for the labels. Also, we do not want a fixed size for all the nodes but rather allow *Graphviz* to adapt the node size to fit the labels. This is controlled by the logical layout parameter `fixedsize`. KEGG laid out the pathway from left to right and we can set this in the graphwide `rankdir` attribute.

```
> attrs = list(node=list(fixedsize=FALSE),
      graph=list(rankdir="LR"))
> width = c(2.5, 1.5, 1.5, 1.5)
> height = c(1.5, 1.5, 1.5, 1.5)
> names(width) = names(height) = long
> nodeAttrs = list(width=width, height=height)
> IMCAGraph = layoutGraph(IMCAGraph, attrs=attrs,
      nodeAttrs=nodeAttrs)
> renderGraph(IMCAGraph)
```

It also makes sense to use a rectangular shape for all but the "Phosphatidylinositol signaling system" node which actually comprises a fairly substantial cellular subprocess and we want it to be highlighted accordingly. The best way to do that is to set the global shape attribute for nodes in `attrs` and adjust for the single "Phosphatidylinositol signaling system" node in `nodeAttrs`.

```
> attrs$node$shape="rectangle"
> shape = "ellipse"
> names(shape) = long[1]
> nodeAttrs$shape = shape
> IMCAGraph = layoutGraph(IMCAGraph, attrs=attrs,
      nodeAttrs=nodeAttrs)
> renderGraph(IMCAGraph)
```

The other three nodes with the long names and also the "F-actin" node represent processes rather than physical objects and we do not want to plot shapes for them, but display plain text of the node names instead (Figure 12.6).

```
> plaintext = rep("transparent", 4)
> names(plaintext) = c(long[2:4], "F-actin")
> nodeRenderInfo(IMCAGraph) = list(fill=plaintext,
      col=plaintext)
> renderGraph(IMCAGraph)
```

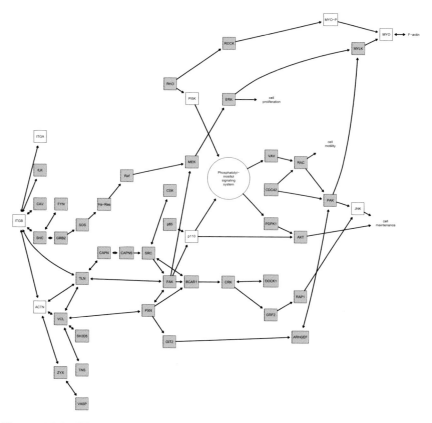

Figure 12.6. The integrin-mediated cell adhesion pathway as rendered by **Rgraphviz** after a considerable amount of fine tuning.

Exercise 12.4
What is still missing in our graph is some color. Looking at Figure 12.4 we can see that there seem to be different classes of nodes, some colored green and others remaining transparent. Reproduce this color scheme for our plot.

12.3.1 Reciprocated edges

There is an option `recipEdges` that determines how reciprocated edges in a graph will be handled. The two options are `combined` (the default) and `distinct`. The `combined` option will display them as a single edge with an arrow on both ends whereas `distinct` shows them as two separate edges. Note that the layout of a graph may change significantly depending on the selection made here. For distinct edges we need more space and most of the layout algorithms are likely to take that extra space into account.

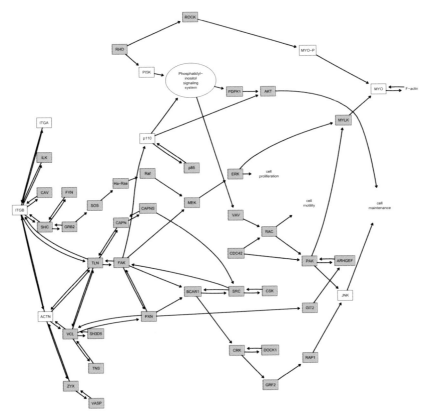

Figure 12.7. The integrin-mediated cell adhesion pathway as rendered by **Rgraphviz** with distinct reciprocated edges.

The following code plots our IMCA graph with distinct reciprocated edges (Figure 12.7):

```
> IMCAGraph = layoutGraph(IMCAGraph, attrs=attrs,
      nodeAttrs=nodeAttrs, recipEdges="distinct")
> renderGraph(IMCAGraph)
```

12.4 Subgraphs

We have considerably improved the presentation of our graph just by changing the node shapes and colors. However, the topological properties of the pathway are still not captured in our output. For instance, there are collections of nodes that represent structures or processes that are known to be closely related. Also, there is some notion of inputs and outputs for the pathway that we would like to show in the layout. The layout algorithm itself does not know about these properties and tries to arrange nodes purely by the edges connecting them. We have to explicitly tell *Graphviz* that we want certain nodes to be kept together and we can do that by means of subgraphs. This concept is distinct from that of subgraphs in the **graph** package.

In the context of graph layout, subgraphs are organizatorial units within a graph and they can be specified using `layoutGraph`'s `subGList` argument which is a list of lists, with each sublist containing one to three of the following elements.

- `graph`: The actual `graph` object for this subgraph. This is the only mandatory element of the inner lists. The easiest way to define this is to use the `subGraph` function from the **graph** package.

- `cluster`: A logical value noting if this is a `cluster` or a `subgraph`. A value of `TRUE` (the default, if this element is missing) indicates a `cluster`. In *Graphviz*, subgraphs are used as an organizational mechanism but are not necessarily laid out in such a way that they are visually together. Clusters are laid out as a separate graph, and thus *Graphviz* will tend to keep nodes of a cluster together.

- `attrs`: A named vector of attributes, where the names are the attribute and the elements are the value. If there are no attributes to specify for this subgraph then `attrs` is unnecessary. For more information about layout attributes, see the documentation of `layoutGraph`

In the following code we specifiy the cell membrane components, the cytoskeleton components, and the affected downstream processes as individual subgraphs. We also add information about inputs and outputs to the pathway through the `sink` and `source` layout attributes.

```
> sg1 = subGraph(c("ITGA", "ITGB", "ILK", "CAV"), IMCAGraph)
> sg2 = subGraph(c("cell maintenance", "cell motility",
        "F-actin", "cell proliferation"), IMCAGraph)
> sg3 = subGraph(c("ACTN", "VCL", "TLN", "PXN", "TNS", "VASP"),
        IMCAGraph)
> subGList = vector(mode="list", length=3)
> subGList[[1]] = list(graph=sg1, attrs=c(rank="source"),
        cluster=TRUE)
> subGList[[2]] = list(graph=sg2, attrs=c(rank="sink"))
> subGList[[3]] = list(graph=sg3)
> IMCAGraph = layoutGraph(IMCAGraph, attrs=attrs,
        nodeAttrs=nodeAttrs, subGList=subGList)
> renderGraph(IMCAGraph)
```

The plot in Figure 12.8 is pretty close to the KEGG image which most likely was produced manually. The integrins and other cell membrane

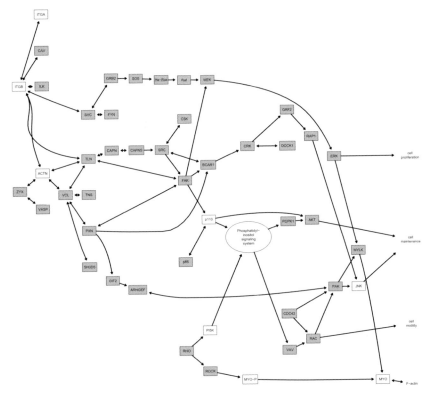

Figure 12.8. Layout of the integrin-mediated cell adhesion pathway using subgraphs to reflect its topology.

components are all lined up on the left side of the plot and the affected downstream processes are located together on the right margin. Actin and the other cytoskeleton components form a nice cluster.

Exercise 12.5
The components of the MAP-kinase signaling pathway ("GRB2", "SOS", "Ras", "Raf", "MEK", "ERK") are still a bit scattered. Use another subgraph to keep them closer together.

12.5   Tooltips and hyperlinks on graphs

The amount of information one can include in a graph plot is limited by the available space on the screen. Interactivity can help overcome this limitation, but R's graphical engine does not offer much interaction. A straightforward way to produce interactive output is to use clickable

HTML. The `imageMap` method for `graph` objects in the **biocGraph** package creates code that can be embedded in a HTML document and that allows for drilldown starting from a static graph image to more detailed information. In this section we show an example using a bipartite graph of co-cited genes in the PubMed literature. Edges connect genes to papers in which they were cited, and clicking on a "paper" node should directly link to the respective abstract on PubMed.

Our example data are in the annotation package **org.Hs.eg.db** which offers mapping from EntrezGene identifiers to PubMed citations and we first have to create a *graph* object from that. We start with one gene (EntrezGene ID 1736) and fetch all papers that cite this gene.

```
> library("org.Hs.eg.db")
> library("biocGraph")
> g1    = "1736"
> paps = org.Hs.egPMID[[g1]]
> genes = mget(paps, org.Hs.egPMID2EG)
> names(genes) = paps
> len = sapply(genes, length)
> table(len)
len
     1      2      3      4      5     58    107    208    211    375
    15      1      2      1      2      1      1      1      1      1
   422    426    503    716    718    773    791   1538   1746   1860
     1      1      1      1      1      1      1      1      1      1
 11328  17988
     1      1
```

There are papers that cite a lot of genes and it is unlikely that they refer to interesting properties of specific genes, so we remove them. Let's remove every paper that mentions more than five genes. To easily distinguish between paper and gene nodes we prepend LL and PM to the node names.

```
> sel = len < 5
> genes = genes[sel]
> paps = paps[sel]
> LLstring = function(i) paste("LL", i, sep=":")
> PMstring = function(i) paste("PM", i, sep=":")
> nd = c(LLstring(unique(unlist(genes))),
        PMstring(paps))
> ed = lapply(nd, function(z) list(edges=integer(0)))
> names(ed) = nd
```

We can use these data now to create the *graph* object `bpg`:

```
> for(i in 1:length(genes)) {
      p = PMstring(names(genes)[i])
      g = LLstring(genes[[i]])
      ed[[p]] = list(edges=match(g, nd))
  }
> bpg = new("graphNEL", nodes=nd, edgeL=ed,
      edgemode="directed")
```

Let's tweak some of the layout and rendering parameters using the tools described earlier to get a more appealing output.

```
> nt = match(substr(nodes(bpg), 1, 2), c("LL", "PM"))
> fill = c("lightblue", "salmon")[nt]
> shape = c("ellipse", "rect")[nt]
> names(fill) = names(shape) = nodes(bpg)
> attrs = list(node=list(fixedsize=TRUE, shape=shape))
> nodeRenderInfo(bpg) = list(fill=fill)
> graph.par(list(nodes=list(fontsize=10)))
```

We have to draw the plot (as shown in Figure 12.9) on a bitmap device stored as a file because this file will later be embedded in the HTML

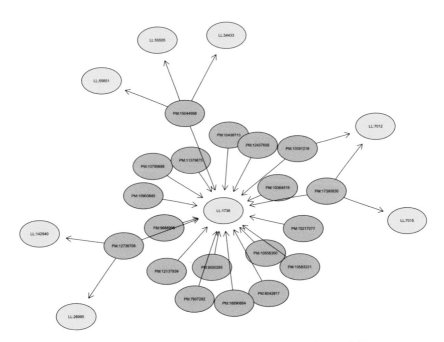

Figure 12.9. A bipartite co-citation graph of PubMed IDs and Entrez genes.

document. Either jpeg or png is good choice for that. For sufficient space for
the graph we have to increase the dimension of the plotting device using
the width and height arguments. Note that renderGraph returns a graph
object with information about the pixel coordinates of each node added
to its renderInfo slot, so we need to make sure that we assign the output back into bgp. The image file will be created in your current working
directory.

```
> imgname = "graph.png"
> png(imgname, width=1024, height=768)
> bpg = layoutGraph(bpg, layoutType="neato", attrs=attrs)
> bpg = renderGraph(bpg)
> dev.off()
```

The imageMap method for graph objects has four mandatory arguments:
object which is the *graph* object, a connection con to the HTML document,
tags which is a named list of HTML tags for links and tooltips, and ima-
gename which is the image file with which the HTML image map will be
associated. You should consult imageMap's documentation for more details.
For setting up the HTML file connection we can use the openHtmlPage
function which will generate the necessary HTML headers.

```
> fhtml = "index.html"
> con = openHtmlPage(fhtml, paste("PubMed co-citations of",
        "gene '1736' Please click on the nodes"))
```

We have already set up the file connection and the *graph* object and we
have plotted it in the image file, so the only thing we still need is the list
of tags. PubMed offers a query syntax to directly retrieve paper abstracts
for valid PubMed identifiers via a URL and we use that as hyperlinks for
our graph:

```
> pnodes = nodes(bpg)[nt==2]
> links = character(length(pnodes))
> tooltips = pnodes
> links =  paste("http://www.ncbi.nih.gov/entrez/query.fcgi",
        "?tool=bioconductor&cmd=Retrieve&db=PubMed&",
        "list_uids=", gsub("PM:", "", pnodes), sep="")
> names(links) = names(tooltips) = pnodes
> tags = list(HREF=links, TITLE=tooltips)
```

Now that we have all the bits and pieces together we can start to assemble
the HTML document

```
> imageMap(bpg, con=con, tags=tags, imgname=imgname)
> closeHtmlPage(con)
```

and have a look at the result by pointing a Web browser to it.

```
> fhtml
> browseURL(fhtml)
```

13

Gene Set Enrichment Analysis

R. Gentleman, M. Morgan, and W. Huber

Abstract

Gene Set Enrichment Analysis (GSEA) is an important method for analyzing gene expression data. It is useful for finding biological themes in gene sets, and it can help to increase the statistical power of analyses by aggregating the signal across groups of related genes. In this chapter, we introduce tools available in the **Category** and **GSEABase** packages for carrying out gene set enrichment analysis.

13.1   Introduction

In this case study, we see how to use gene set enrichment analysis (GSEA); Subramanian et al. (2005); Tian et al. (2005); Jiang and Gentleman (2007). We primarily concentrate on the two-sample problem, where the data can be divided into two distinct groups, and we want to understand the set of differentially expressed genes between the two groups. We use a subset of the **ALL** dataset as described in Chapter 1.

The basic idea behind GSEA is to use predefined sets of genes, usually derived from functional annotation or from results of prior experiments, in order to better interpret the experiment that we are analyzing. The idea is similar to the one of Hypergeometric testing of $2x2$ contingency tables (see Chapter 14 for more details on that approach). Perhaps the most important difference is that in GSEA there is no need to make a hard cutoff between genes that are differentially expressed and those that are not. Rather, we determine a continuous-valued score, for example, the t-statistic, and see whether its values are associated with the gene sets of interest.

Any collection of gene sets can be used, and in many cases users will avail themselves of predefined gene sets, such as those available from GOA or KEGG. In this chapter we concentrate on KEGG, chromosome bands, and protein domains, because many other examples have been presented of applying these procedures to GO annotations.

F. Hahne et al., *Bioconductor Case Studies*, DOI: 10.1007/978-0-387-77240-0_13,
© Springer Science+Business Media, LLC 2008

In release 2.1 of Bioconductor the **GSEABase** package was introduced. It contains software infrastructure for performing GSEA analyses. The basic concepts are simple: the *GeneSet* class represents a set of genes that are all described by a common set of identifiers such as EntrezGene IDs. A *GeneSetCollection* object is a collection of gene sets. **GSEABase** provides tools for performing set operations on gene sets, such as union and intersection, as well as ways to construct specific gene set collections, such as those for GO and KEGG, and using different types of inputs, such as a Bioconductor annotation package or an *ExpressionSet*.

There are a number of other Bioconductor packages that provide GSEA-type analyses. They include **PGSEA**, **sigPathways**, and **GlobalAncova**. For the Hypergeometric testing approach, the packages **GOstats** and **topGO** provide specialized methods that try to remove redundancy by taking advantage of the nested structure of the Gene Ontologies (The Gene Ontology Consortium, 2000).

13.1.1 Simple GSEA

Prior to conducting a GSEA analysis, we recommend making a basic data quality assessment followed by filtering genes to remove those that do not show much variation across samples (and hence have little ability to discriminate any samples). It is essential that sample annotation, such as phenotypic characteristics, not be used in this filtering step.

Next we compute a test statistic for each remaining gene, possibly using some form of moderation or regularization with an empirical Bayes method. It does not really matter how this step is carried out, but it is important that you make a choice that you are comfortable with, and that the resulting quantity have a similar distribution, and interpretation, for all genes being used. In particular, one should avoid statistics that have a different baseline for each gene, for example, statistics that are proportional to the observed intensity. In the following examples, which consider a two-group comparison, we use the *t*-statistic.

Now, the basic idea is that under the null hypothesis of no difference in mean expression between the two groups, the per-gene *t*-statistics t_k have a *t*-distribution. If we further assume that the genes, and hence the observed *t*-statistics, are independent, their sum, taken over the genes in a gene set, divided by the square root of the number of genes, should have an approximately Normal distribution with mean zero and variance one, by the central limit theorem. So that with

$$z_K = \frac{1}{\sqrt{|K|}} \sum_{k \in K} t_k, \tag{13.1}$$

where K denotes the gene set, and $|K|$ the number of genes in the gene set, the z_k have approximately a standard Normal distribution. And so, one can potentially identify interesting gene sets by comparing their z_K to the quantiles of a standard Normal distribution. The assumption that the the genes are independent is quite strong, but in practice the test seems to lead to reasonable results.

An alternative approach is to make use of a permutation test to assess which gene sets have an unusually large absolute value of z_K. The usual null hypothesis is that the sample labels are not related to the observed values of gene expression, and hence by permuting labels we can generate a reference distribution for any test statistic of interest. When there are relatively few observations one will typically compute all possible permutations, but even for modest sample sizes it is not practical to compute all permutations and instead we sample some large number (typically 1000) of permutations. The observed value of the z_K-statistic is then compared to the reference distribution to obtain a p-value.

13.1.2   Visualization

The use of graphical tools can help to better understand how well the data are being modeled and they provide diagnostic checks on various assumptions you might have made.

A Q–Q plot of the observed test statistics (or p-values) versus an appropriate reference distribution will help to visually identify how extreme some of the per gene set statistics are. Once specific gene sets are identified as being of interest, heatmaps of the gene set data can be very informative. The functions KEGG2heatmap and GO2heatmap can easily be extended to other situations.

For two-sample comparisons we use plots that show the mean expression value in each group. Two specialized functions are the GOmnplot and KEG-Gmnplot. It would also be useful to produce side-by-side boxplots on a per gene basis in situations where there are sufficient samples.

13.1.3   Data representation

The **GSEABase** package provides basic software tools for dealing with gene sets. A gene set is a set of genes that someone (possibly you) has determined are of interest. These gene sets can be grouped together into collections, and most analyses will be performed on collections. **GSEABase** provides tools for performing the usual set operations such as unions and intersection on gene sets.

A gene set in the *GeneSet* class is represented, essentially, as a *character* vector of identifiers together with information denoting the identifier system to which the identifiers refer. An alternative representation, which may be especially useful in the context of gene set collections, is as an incidence matrix, where, say, the rows correspond to the different gene sets, and there is one column for each gene. The entries in the matrix $A[i, j]$ are either 1 or 0, depending on whether gene j is in gene set i. The incidence matrix is very useful for many of the computations we want to carry out and is used in most of the examples. The `incidence` function from the **GSEABase** package can be used to compute the incidence matrix from a gene set collection. Simple operations such as computing column or row sums will tell you how many gene sets a gene is in, or how many genes are in a particular gene set.

When an analysis identifies multiple gene sets as significant, we have found it valuable to characterize the amount of overlap between them. This can easily be computed from the incidence matrix. Gene sets that overlap substantially may be complementary, or one may be able to determine that the effect seems to be due to only one of the gene sets. Some examples and more explicit discussion were given by Jiang and Gentleman (2007).

13.2   Data analysis

We begin by describing the preprocessing steps that are used on the data, and follow that with applications using KEGG pathways and chromosome location. We note that the use of chromosome location is sensible for these data, as they are derived from cancer cells, and it is well known that some of the genomic alterations that are related to cancer tend to cluster by genomic location, perhaps due to amplification or deletion of DNA, or due to methylation or demethylation events.

13.2.1   Preprocessing

For this chapter, we use the `ALL` data, which have been obtained in a microarray study of B- and T-cell leukemia. We want to find genes that are differentially expressed between two distinct types of B-cell leukemia.

```
> library("ALL")
> data("ALL")
```

The data and the following steps with which we construct the subset of interest, `ALL_bcrneg`, are described in more detail in Chapter 1. Briefly, we select samples with B-cell leukemia harboring the BCR/ABL translocation and those samples with no observed cytogenetic abnormalities (NEG).

```
> bcell = grep("^B", as.character(ALL$BT))
> moltyp = which(as.character(ALL$mol.biol)
      %in% c("NEG", "BCR/ABL"))
> ALL_bcrneg = ALL[, intersect(bcell, moltyp)]
> ALL_bcrneg$mol.biol = factor(ALL_bcrneg$mol.biol)
```

The last line in the code above is used to drop unused levels of the *factor* variable `mol.biol`. Nonspecific filtering removes the probe sets that we believe are not sufficiently informative, so that there is little point in considering them further. We use the function `nsFilter` from the **genefilter** package to apply a number of different criteria. For instance, by default the function removes the control probes on Affymetrix arrays, which can be identified by their `AFFX` prefix. It also excludes probe sets without Entrez-Gene ID, and for instances of multiple probe sets mapping to the same EntrezGene ID, only the probe set with the largest variability is retained. The choice of the `var.cutoff` parameter indicates that we select the top 50% of probe sets on the basis of variability.

```
> library("genefilter")
> ALLfilt_bcrneg = nsFilter(ALL_bcrneg, var.cutoff=0.5)$eset
```

Exercise 13.1

　　a　*How many samples are in our subset? How many are BCR/ABL and how many NEG?*

　　b　*How many probe sets have been selected for our analysis?*

13.2.2　Using KEGG

KEGG (Kanehisa and Goto, 2000) provides mappings of genes to pathways and this information is included in most Bioconductor annotation packages. Many investigators are interested in whether there is some indication that certain pathways are implicated in their analysis. One way to make that assessment is to perform a gene set analysis on the KEGG pathways.

We use code from the **GSEABase** package to compute the incidence matrix that maps between probes and the pathways. The function `GeneSet-Collection` can produce gene sets for a number of different inputs, including an instance of the *ExpressionSet* class, which is what we use.

```
> library("GSEABase")
> gsc = GeneSetCollection(ALLfilt_bcrneg,
                setType=KEGGCollection())
> Am = incidence(gsc)
> dim(Am)
[1]   194 1666
```

We next compute a reduced *ExpressionSet* object `nsF` that retains only those features (genes) that are mentioned in the incidence matrix `Am` and whose features are in the same order as the columns of `Am`.

```
> nsF = ALLfilt_bcrneg[colnames(Am),]
```

Exercise 13.2
How many gene sets and how many genes are represented by the incidence matrix? How many gene sets have fewer than ten genes in them? What is the largest number of gene sets in which a gene can be found?

Next we compute the per gene test statistics using the `rowttests` function from the **genefilter** package. There are several other functions in **genefilter** for performing fast rowwise statistics (e.g., `rowFtests`).

```
> rtt = rowttests(nsF, "mol.biol")
> rttStat = rtt$statistic
```

Exercise 13.3
How many test statistics are positive? How many are negative? How many have a p-value less than 0.01?

In the next code segment, we reduce the incidence matrix by removing all gene sets that have fewer than ten genes in them.

```
> selectedRows = (rowSums(Am)>10)
> Am2 = Am[selectedRows, ]
```

Finding general trends requires that you use gene sets with a reasonable number of genes, and here we have operationalized that by setting our cutoff at ten. This cutoff is arbitrary, and in any analysis you should think about whether to do this, and if so, what value might be used on the basis of what you are interested in finding.

Now it is fairly easy to compute the per gene set test statistics and to produce a Normal Q–Q plot; see Figure 13.1.

```
> tA = as.vector(Am2 %*% rttStat)
> tAadj = tA/sqrt(rowSums(Am2))
> names(tA) = names(tAadj) = rownames(Am2)
```

```
> qqnorm(tAadj)
```

We see that there is one pathway that has a remarkably low observed value (less than −5) so we take a closer look at this pathway.

```
> library("KEGG.db")
> smPW = tAadj[tAadj < (-5)]
> pwName = KEGGPATHID2NAME[[names(smPW)]]
> pwName
[1] "Ribosome"
```

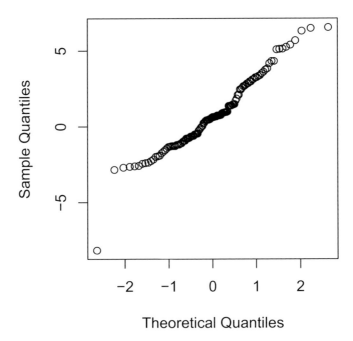

Figure 13.1. The per gene set Q–Q plot.

Now we can produce some summary plots based on the genes annotated at this pathway. The mean plot presents a comparison of the average
expression value for each of our two groups, for each gene in the specified
pathway.

```
> KEGGmnplot(names(smPW), nsF, "hgu95av2", nsF$"mol.biol",
    pch=16, col="darkblue")
```

That is, each point in the left panel of Figure 13.2 represents one gene
and the value on the x-axis is the mean in the BCR/ABL samples whereas
the value on the y-axis is the mean value in the NEG samples.

Exercise 13.4
*Many of the points in Figure 13.2 appear to lie above the diagonal. Is this
to be expected?"*

And finally we can produce a heatmap for the genes in the ribosome
pathway. We use the `KEGG2heatmap` function to do most of the hard work,
and the result is shown in the right panel of Figure 13.2.

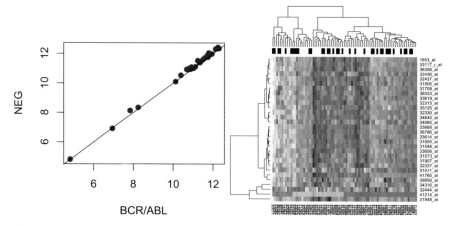

Figure 13.2. The left panel shows the scatterplot of within-group means for the genes in the ribosome pathway. The right panel shows a heatmap for these genes. The black and white bars at the top indicate the two disease types, BCR/ABL and NEG.

```
> sel = as.integer(nsF$mol.biol)
> KEGG2heatmap(names(smPW), nsF, "hgu95av2",
    col=colorRampPalette(c("white", "darkblue"))(256),
    ColSideColors=c("black", "white")[sel])
```

Exercise 13.5
What sorts of things do you notice in the heatmap? The gene labeled 41214_at *has a very distinct pattern of expression. Can you guess what is happening? Hint: look at which chromosome it is on.*

As a further exercise, produce corresponding plots for the pathway with the largest positive average *t*-statistic.

13.2.3 Permutation testing

The assumptions on which we based the test above are somewhat strong, and it is of some interest to consider alternative approaches. For GSEA it is straightforward to compute a permutation test. The gseattperm function in the **Category** package can be used to compute the permutation test. It takes as inputs the gene expression data, phenotypic data for the samples, and the incidence matrix representing the gene sets of interest. The value returned by gseattperm is a matrix with columns Lower and Upper. For each row (gene set) the Lower column gives the proportion of permutation *t*-statistics that were smaller than the observed *t*; the Upper column gives the proportion of the permutation *t*-statistics that were larger than the observed *t*-statistic.

In the code chunk below we compute the permutation distribution based on 1000 permutations. We use a p-value cutoff of 0.025 that corresponds to a two-sided hypothesis test at the 0.05 level. Because we are carrying out a permutation test and want the results to be reproducible we begin by setting the seed for the random number generator. For users that are going to use permutation distributions, we recommend familiarizing themselves with the different pseudo-random number generators in R and the role of the seed, and using these tools with care.

```
> set.seed(123)
> NPERM = 1000
> pvals = gseattperm(nsF, nsF$mol.biol, Am2, NPERM)
> pvalCut = 0.025
> lowC  = names(which(pvals[, 1]<=pvalCut))
> highC = names(which(pvals[, 2]<=pvalCut))
```

In the next code chunk we print some of the resulting pathway names. There is one in lowC and 21 in highC.

```
> getPathNames(lowC)
```

```
[1] "03010: Ribosome"
```

```
> getPathNames(highC)
```

```
[1] "04360: Axon guidance"
[2] "05130: Pathogenic Escherichia coli infection - EHEC"
[3] "05131: Pathogenic Escherichia coli infection - EPEC"
[4] "04520: Adherens junction"
[5] "04510: Focal adhesion"
```

Exercise 13.6
What permutation-base p-value is the most extreme? What does the heatmap look like for this gene set?

Exercise 13.7
Compare the p-values from the parametric analysis to those from the permutation analysis.

13.2.4 Chromosome bands

Another interesting application is to use the mapping of genes to chromosome bands as the gene sets. In doing this, you are studying whether there are anomalies in the pattern of gene expression that relate to chromosomal location. In many cancers there are underlying genomic changes, such as amplifications or deletions that cause coordinated changes in gene expression as a function of chromosomal location. In other settings, a separate justification for using chromosomal location would be needed. The chromosome band information can be obtained from the metadata variable with the suffix MAP, so for us it is hgu95av2MAP.

Exercise 13.8
What does the manual page say is the interpretation of the MAP position 17p33.2*?*

Exercise 13.9
Just as we did for the KEGG analysis, we need to remove probe sets that have no chromosome band annotation. Follow the approach used in the KEGG analysis and create a new ExpressionSet object nsF2*. We will use it later in this chapter.*

Relevant MAP positions can be computed using the MAPAmat function from the **Category** package.

```
> EGtable = toTable(hgu95av2ENTREZID[featureNames(nsF2)])
> entrezUniv = unique(EGtable$gene_id)
> chrMat = MAPAmat("hgu95av2", univ=entrezUniv)
> rSchr = rowSums(chrMat)
```

Exercise 13.10
How many genes were selected? How many map positions?

Exercise 13.11
Further reduce chrMat *so that only bands with at least five genes are retained. Produce a Q–Q plot and identify the interesting bands. Use Google or some other search engine to determine what might be interesting about these bands.*
Do an analysis similar to the one we performed on the KEGG pathways. Produce mean plots, heatmaps, and so on. Try to identify a set of interesting bands.

Exercise 13.12
Reorder the columns of chrMat *so that they are in the same order as the corresponding features in the ExpressionSet object* nsF2*.*

13.3   Identifying and assessing the effects of overlapping gene sets

Once the potentially interesting gene sets have been identified, it is useful to consider how much overlap there is between them. In some cases the list can be reduced by identifying gene sets that are redundant, or that perhaps have been identified because they overlap with another gene set that really is important. It is possible to consider higher levels of overlap, say between three gene sets, but here we restrict our attention to two gene sets at a time.

To assess the amount of overlap we suggest using a simple statistic which is the number of genes in common, divided by the number of genes in the smaller of the two gene sets. This statistic is 0 when there are no genes in common and 1 if one of the two gene sets is a subset of the other. Computation can be performed using the methods in the **GSEABase** package, but is easy enough using generic algebra on the incidence matrix.

We consider the permutation analysis reported in Section 13.2.3 and consider all gene sets with permutation p-values less than 0.025 as interesting. To compute our index, we simply take the submatrix of the incidence matrix that corresponds to the interesting gene sets (as identified earlier) and then take the matrix product with its transpose. The diagonal elements of Amx are the sizes of each gene set, and the off-diagonal elements are the number of genes in common.

```
> Ams = Am2[union(lowC, highC),]
> Amx = Ams %*% t(Ams)
> minS = outer(diag(Amx), diag(Amx), pmin)
> overlapIndex = Amx/minS
```

The resulting matrix, overlapIndex, is plotted in Figure 13.3.

Exercise 13.13
How many genes are in each of the four pathways, 04512, 04940, 04510, 04514? How many are in the overlap for each pair?

In order to determine whether there is evidence that both of the gene sets in a pair are involved in the observed relationship with the t-statistic, or whether one of the two is perhaps only implicated due to the shared genes, we consider a number of different linear models.

First we fit a linear model to each of the two gene sets separately, and observe that the corresponding p-values are less than 0.05. But when both are included in the model, only one remains significant.

```
> P04512 = Ams["04512",]
> P04510 = Ams["04510",]
```

overlapIndex

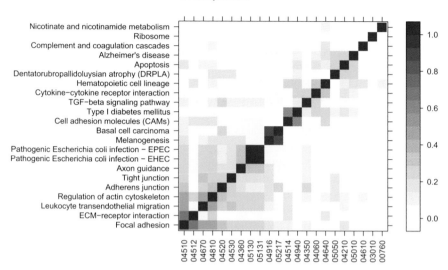

Figure 13.3. Overlap between the gene sets as measured by `overlapIndex`.

```
> lm1 = lm(rttStat ~ P04512)
> summary(lm1)$coefficients
            Estimate Std. Error t value Pr(>|t|)
(Intercept)   0.0948     0.0378    2.51   0.0123
P04512        0.5783     0.2774    2.08   0.0372
> lm2 = lm(rttStat ~ P04510)
> summary(lm2)$coefficients
            Estimate Std. Error t value Pr(>|t|)
(Intercept)   0.0715     0.0385    1.86 0.063565
P04510        0.5913     0.1604    3.69 0.000235
> lm3 = lm(rttStat ~ P04510+P04512)
> summary(lm3)$coefficients
            Estimate Std. Error t value Pr(>|t|)
(Intercept)    0.070     0.0386   1.815  0.06971
P04510         0.541     0.1724   3.139  0.00173
P04512         0.237     0.2973   0.796  0.42627
```

We next divide the genes into three groups: those that are in 04512 only, those that are in both sets, and those that are in 04510 only. Using these variables in our linear regression shows that only the second and third variables are significant. This suggests that the pathway 04512 only appears to be interesting because of the genes it shares with the 04510 pathway.

```
> P04512.Only = ifelse(P04512 != 0 & P04510 == 0, 1, 0)
> P04510.Only = ifelse(P04512 == 0 & P04510 != 0, 1, 0)
> Both        = ifelse(P04512 != 0 & P04510 != 0, 1, 0)
> lm4 = lm(rttStat ~ P04510.Only + P04512.Only + Both)
> summary(lm4)
Call:
lm(formula = rttStat ~ P04510.Only + P04512.Only + Both)

Residuals:
   Min     1Q Median     3Q    Max
-4.167 -1.034 -0.183  0.884  7.211

Coefficients:
              Estimate Std. Error t value Pr(>|t|)
(Intercept)    0.0689     0.0386    1.78   0.0749 .
P04510.Only    0.5650     0.1804    3.13   0.0018 **
P04512.Only    0.4088     0.4842    0.84   0.3985
Both           0.6972     0.3353    2.08   0.0377 *
---
Signif. codes:  0 '***' 0.001 '**' 0.01 '*' 0.05 '.' 0.1
 ' ' 1

Residual standard error: 1.53 on 1662 degrees of freedom
Multiple R-Squared: 0.0086,    Adjusted R-squared: 0.00
681
F-statistic: 4.81 on 3 and 1662 DF,  p-value: 0.00245
```

We also recommend checking the sign of the t-statistic, to make sure that effects are concordant (as they are here).

Exercise 13.14
Repeat this analysis for the other pair of pathways, 04940 and 04514.

14

Hypergeometric Testing Used for Gene Set Enrichment Analysis

S. Falcon and R. Gentleman

Abstract

After the set of interesting genes has been determined, say those that are differentially expressed, a next step in the analysis is to attempt to find functional relationships among those genes that might help better elucidate the underlying biology. These methods typically rely on existing or predefined sets of genes. In this chapter we show how to carry out Hypergeometric tests to identify potentially interesting gene sets.

14.1   Introduction

The **Category** and **GOstats** packages provide extensive facilities for the testing of over- and underrepresentation of gene sets in a specified list of interesting genes. In this chapter we focus most of our examples on the gene set collection induced by the Gene Ontology GO; (The Gene Ontology Consortium, 2000). However, the techniques demonstrated can be easily translated for use with other gene set collections supported by the **Category** package including KEGG, PFAM, and chromosome band annotation and these are covered in the exercises.

In this chapter we describe the preprocessing required to construct inputs for the main testing function, `hyperGTest`, the algorithms used, and the structure of the return value. We use a microarray data set (Chiaretti et al., 2004, 2005) from a clinical trial in acute lymphoblastic leukemia (ALL) to work an example analysis. In the ALL data, we focus on the patients with B-cell derived ALL, and in particular on comparing the group with BCR/ABL translocations to those with no observed cytogenetic abnormalities (NEG).

F. Hahne et al., *Bioconductor Case Studies*, DOI: 10.1007/978-0-387-77240-0_14,
© Springer Science+Business Media, LLC 2008

To get started, load the packages needed for this analysis:

```
> library("Biobase")
> library("ALL")
> library("hgu95av2.db")
> library("annotate")
> library("genefilter")
> library("GOstats")
> library("RColorBrewer")
> library("xtable")
> library("Rgraphviz")
```

14.2   The basic problem

The reasoning behind this approach is that there is some universe of objects (in our case genes) that are of potential interest, and that these objects can be divided into two groups (those that are interesting and those that are not). In addition, there are other characteristics of the objects that are also binary, such as belonging to a particular GO category, or having a particular biological property. And hence one would like to ask whether there is an association between being interesting and having the particular property. This question is easily answered using basic probability, and the resulting test is also widely known as Fisher's exact test.

The probability calculation can be carried out in two different ways, but the resulting test statistics and p-values are identical. Consider an urn containing one ball for each gene in the universe and imagine that those that are interesting are colored black, and those that are not interesting are colored white. Then, under the null hypothesis that there is no relationship between being interesting and being in a given GO category containing K genes, we can model the number of interesting genes using a Hypergeometric distribution. If there are j interesting genes in the GO category, we simply compute the probability of seeing j or more black balls in K draws, without replacement, from the urn. This probability is symmetric, in the sense that we could also have described the problem with the balls colored according to the GO category, and select, without replacement, one ball for each gene in our gene list. Another way of thinking about this is to draw the balls from the urn that represent genes annotated at the given term and fill out a two-way table as shown in Table 14.1.

In principle there is no reason why either grouping needs to be binary. You could have three types of genes (really interesting, sort of interesting, and not interesting) and a category that has three levels. If so, the multivariate generalizations of the Hypergeometric distribution will be needed.

Table 14.1. The two-way table for testing overrepresentation of a GO term among a selected list of interesting genes.

	Interesting (Black)	Not (White)
In GO term	n_{11}	n_{12}
Not in GO term	n_{21}	n_{22}

Selection of the universe is very important; making it too large, or too small, will have a large impact on the observed p-values. Our recommendation is to include in the universe only those genes that could have been selected as interesting. Practically speaking, for a microarray experiment that means that the universe consists of all genes that are represented by probes on the array. In other settings one may want to use all genes from an organism, but you should be very cautious in doing this, it will make your p-values more extreme, and suggest relationships where none exist.

A second practical issue that arises in the analysis of microarray experiments is that often some genes are represented by more than one probe on the array. This is also problematic, and in order for the Hypergeometric probabilities to be correct, you must restrict yourself to a single value for each gene. You could take a different approach, and develop the correct probability model to deal with multiple probes per gene, but the software in the **GOstats** and **Category** packages should not be used directly for those applications.

Some issues naturally arise, such as multiple testing. You will be carrying out hundreds, if not thousands, of tests and one might wonder if appropriate p-value correction methods cannot be found. It seems that this is difficult, and our approach here is described in Falcon and Gentleman (2007) which is similar to that of Alexa et al. (2006). And, you may also want to consider the gene set enrichment analysis (GSEA)[gene set enrichment analyis] (Subramanian et al., 2005; Tian et al., 2005; Jiang and Gentleman, 2007) as an alternative.

14.3   Preprocessing and inputs

We propose a template that can be used for any analysis in which one wishes to determine if a gene set from a collection of gene sets is over- or underrepresented relative to a specified list of genes. The template consists of the following steps.

1. Perform nonspecific filtering.

2. Define the gene universe.

3. Determine a subset of interesting genes.

4. Test for over- or underrepresentation in the collection of gene sets.

To perform an analysis using the Hypergeometric-based tests, one needs to define a *gene universe* (the balls in the urn) and a list of interesting genes from the universe. Although it is clear that the selected gene list determines to a large degree the results of the analysis, the fact that the universe has a large effect on the conclusions is, perhaps, less obvious. It is worth noting that the effect of increasing the universe size with genes that are irrelevant to the questions at hand, in general, has the effect of making the p-values look more significant. For example, in a universe of 1000 genes where 400 have been selected, suppose that a GO term has 40 gene annotations from the universe of 1000. If 10 of the genes in the interesting gene list are among the 40 genes annotated in this category, then the Hypergeometric p-value is 0.99. However, if the gene universe contained 5000 genes, the p-value would drop to 0.001.

For microarray data, one can use the unique gene identifiers assayed in the experiment as the gene universe. However, the presence of a gene on the array does not necessarily mean much. Some arrays, such as those from Affymetrix, attempt to include probes for as much of the genome as possible. Because not all genes will be expressed under all conditions (a widely held belief is that about 40% of the genome is expressed in any tissue), it may be sensible to reduce the universe to those that are expressed.

To identify the set of expressed genes from a microarray experiment, we propose that a nonspecific filter be applied and that the genes that pass the filter be used to form the universe for any subsequent functional analyses. Below, we outline the nonspecific filtering procedure used for the example analysis.

Once a gene universe has been established, one can apply any number of methods to select genes. For the example analysis we use a simple t-test to identify differentially expressed genes among the two subgroups in the sample population.

14.3.1 Nonspecific filtering

First we load the ALL data object and extract the subset of the data we wish to analyze: subjects with either no cytogenetic abnormality (NEG) or those harboring BCR/ABL translocations. In Chapter 1 you can find a more detailed explanation about the **ALL** data and the individual steps of the subsetting.

```
> data(ALL)
> bcell = grep("^B", as.character(ALL$BT))
> types = c("NEG", "BCR/ABL")
> moltyp = which(as.character(ALL$mol.biol) %in% types)
> ALL_bcrneg = ALL[, intersect(bcell, moltyp)]
> ALL_bcrneg$mol.biol = factor(ALL_bcrneg$mol.biol)
```

Exercise 14.1

   a. *How many samples are in the subset? How many have the BCR/ABL phenotype?*

   b. *What is the name of the annotation package associated with these data? How many probe sets are represented in the dataset?*

The **genefilter** package provides the `nsFilter` function that makes it easy to apply our standard nonspecific filter to an *ExpressionSet*. In the call to `nsFilter` shown below, we indicate that probe sets must have an EntrezGene ID and an annotation in the GO BP ontology. We've also asked to filter out lowvariance probe sets using the `IQR` function with a cutoff of 0.5 and to remove features with an ID beginning with the string "AFFX".

```
> varCut = 0.5
> filt_bcrneg = nsFilter(ALL_bcrneg, require.entrez=TRUE,
      require.GOBP=TRUE, remove.dupEntrez=TRUE,
      var.func=IQR, var.cutoff=varCut,
      feature.exclude="^AFFX")
> names(filt_bcrneg)
[1] "eset"        "filter.log"
> ALLfilt_bcrneg = filt_bcrneg$eset
```

Exercise 14.2
What does `nsFilter`*'s* `remove.dupEntrez` *argument do?*

Exercise 14.3
How many probe sets were removed because they duplicated a mapping to an EntrezGene ID?

Your nonspecific filtering needs may not be met by the filtering options provided by the `nsFilter` function. For example, because there is an imbalance of men and women by group, we continue the filtering by removing probe sets that measure genes on the Y chromosome.

Exercise 14.4
Remove probe sets that measure genes located on the Y chromosome. (Hint: use the `hgu95av2CHR` *environment.)*

We also remove any probe sets that do not have a gene symbol annotation.

```
> hasSymbol = sapply(mget(featureNames(ALLfilt_bcrneg),
        envir=hgu95av2SYMBOL), function(x)
        !(length(x) == 1 && is.na(x)))
> ALLfilt_bcrneg = ALLfilt_bcrneg[hasSymbol, ]
```

The gene identifiers corresponding to the probe sets that remain after the nonspecific filtering define the gene universe we use for the Hypergeometric tests.

```
> affyUniverse = featureNames(ALLfilt_bcrneg)
> entrezUniverse = unlist(mget(affyUniverse,
        hgu95av2ENTREZID))
```

Exercise 14.5
Define an alternate gene universe based on the entire microarray chip.

Summary of nonspecific filtering

Our nonspecific filtering procedure removed probes missing EntrezGene or Symbol identifiers as well as those lacking a mapping to at least one GO term. The interquartile range was used with a cutoff of 0.5 to select probes with sufficient variability across samples to be informative; probes with little variability across all samples are inherently uninteresting. The set of remaining probes was refined by ensuring that no two probe sets map to the same EntrezGene identifier. For those probes mapping to the same EntrezGene ID, the probe with largest the IQR was selected. Because of an imbalance of men and women by group, probes measuring genes on the Y chromosome were dropped. Finally, we removed probe sets missing a gene symbol annotation.

Producing a set of EntrezGene identifiers that map to a unique set of probes at the nonspecific filtering stage is important because genes are mapped to GO categories using EntrezGene IDs and we want to avoid double-counting any GO categories. In all, the filtering left 4229 genes.

14.3.2   Gene selection via t-test

We apply a standard *t*-test to identify a set of genes with differential expression between the BCR/ABL and NEG groups.

```
> ttestCutoff = 0.05
> ttests = rowttests(ALLfilt_bcrneg, "mol.biol")
> smPV = ttests$p.value < ttestCutoff
> pvalFiltered = ALLfilt_bcrneg[smPV, ]
> selectedEntrezIds = unlist(mget(featureNames(pvalFiltered),
        hgu95av2ENTREZID))
```

Exercise 14.6
How many probe sets have a p-value less than the cutoff?

We did not make use of any *p*-value correction methods, or modified *t*-statistics, because our focus is on Hypergeometric testing. But you could make use of the capabilities of the **limma** package should you desire.

A detail often omitted from gene set association analyses is the fact that the *t*-test, and most similar statistics, are directional. For a given gene, average expression might be higher in the BCR/ABL group than in the NEG group, whereas for a different gene it might be the NEG group that shows the increased expression. By only looking at the *p*-values for the test statistics, the directionality is lost. The danger is that an association with a gene set may be found where the genes are not differentially expressed in the same direction. One way to tackle this problem is by separating the selected gene list into two lists according to direction and running two analyses. A more elegant approach is the subject of further research.

14.3.3   Inputs

Often one wishes to perform many similar analyses using slightly different sets of parameters. To facilitate this, the main interface to the Hypergeometric tests, `hyperGTest`, takes a single parameter object as its argument. This argument is a *GOHyperGParams* instance. There are also parameter classes *KEGGHyperGParams* and *PFAMHyperGParams* defined in the **Category** package that allow for testing for association with KEGG pathways and PFAM protein domains, respectively.

Using a parameter class instead of individual arguments makes it easier to organize and execute a series of related analyses. For example, one can create a list of *GOHyperGParams* instances and perform the Hypergeometric test on each using R's `lapply` function.

In the absence of a parameter class, this could be achieved using `mapply`, but the result would be less readable. Because parameter objects can be copied and modified, they tend to reduce duplication of code. We demonstrate this in the following example.

Below, we create a parameter instance by specifying the gene list, the universe, the name of the annotation data package, and the GO ontology we wish to interrogate. For the example analysis, we have stored the vector of EntrezGene identifiers making up the gene universe in `entrezUniverse`. The interesting genes are stored in `selectedEntrezIds`. If you are following along with your own data and have an *ExpressionSet* instance resulting from a nonspecific filtering procedure, you can create the `entrezUniverse` and `selectedEntrezIds` vectors using code similar to that shown below. Be sure to verify that no gene IDs are duplicated in the universe set.

```
> ## if you are following along with your own data...
> entrezUniverse = unlist(mget(featureNames(yourData),
      hgu95av2ENTREZID))
> pvalFiltered = yourData[hasSmallPvalue, ]
> selectedEntrezIds = unlist(mget(featureNames(pvalFiltered),
      hgu95av2ENTREZID))
```

Here is a description of all the arguments needed to construct a *GOHyperGParams* instance.

geneIds: A vector of gene identifiers that defines the selected list of genes. This is often the output of a test for differential expression among two sample groups. For most expression arrays, this will be a vector of EntrezGene IDs. If you are using the **YEAST** annotation package, the vector will contain systematic names.

universeGeneIds: A vector of gene identifiers that defines the universe of possible genes. We recommend using the set of gene IDs that result from nonspecific filtering. The identifiers should be of the same type as the geneIds.

annotation: A string giving the name of the annotation data package that corresponds to the chip used in the experiment.

ontology: A two-letter string specifying one of the three GO ontologies: BP, CC, or MF. The hyperGTest function only tests a single GO ontology at one time.

pvalueCutoff: A numeric value between zero and one used as a cutoff for *p*-values generated by the Hypergeometric test. When the test being performed is nonconditional, this is only used as a default value for printing and summarizing the results. For a conditional analysis, the cutoff is used during the computation to perform the conditioning: child terms with a *p*-value less than pvalueCutoff are conditioned out of the test for their parent term.

conditional: A logical value. If TRUE, the test performed uses the conditional algorithm. Otherwise, a standard Hypergeometric test is performed.

testDirection: A string that can be either "over" or "under". This determines whether the test performed detects over- or underrepresented GO terms.

```
> hgCutoff = 0.001
> params = new("GOHyperGParams", geneIds=selectedEntrezIds,
      universeGeneIds=entrezUniverse, annotation="hgu95av2.db",
      ontology="BP", pvalueCutoff=hgCutoff, conditional=FALSE,
      testDirection="over")
```

14.4   Outputs and result summarization

14.4.1   Calling the hyperGTest function

The hyperGTest function returns an instance of class *GOHyperGResult*.
When the input parameter object is a *KEGGHyperGParams* or *PFAMHy-perGParams* instance, the result will instead be a *HyperGResult* object (the
GO case is special because of the relationship among GO terms). Most of
the reporting and summarization methods demonstrated here will work the
same, except for those that deal specifically with GO or the GO graph.

As shown below, printing the result at the R prompt provides a brief
summary of the test performed and the number of significant terms found.

```
> hgOver = hyperGTest(params)


> hgOver
Gene to GO BP  test for over-representation
1841 GO BP ids tested (6 have p < 0.001)
Selected gene set size: 713
    Gene universe size: 4229
    Annotation package: hgu95av2
```

14.4.2   Summarizing a GOHyperGResult object

The summary function returns a *data.frame* summarizing the result
(Table 14.2). By default, only the results for terms with a *p*-value less than
the cutoff specified in the parameter instance will be returned. However,
you can set a new cutoff using the pvalue argument. You can also set a
minimum number of genes for each term using the categorySize argument.
For *GOHyperGResult* objects, the summary method also has a htmlLinks
argument. When TRUE, the GO term names are printed as HTML links to
the GO Web site.

Exercise 14.7
  a. *What columns are included in the data.frame returned by the* summary
     method when called on a GOHyperGResult instance?

Table 14.2. Top six overrepresented GO terms with at least 100 gene annotations identified by hyperGTest.

	Count	Size	Term
GO:0007154	241	1170	cell communication
GO:0007165	225	1086	signal transduction
GO:0006955	65	257	immune response
GO:0022610	57	209	biological adhesion
GO:0007155	57	209	cell adhesion
GO:0009966	51	182	regulation of signal transduct

b. *How many GO IDs have a p-value less than 0.05 and include at least 350 genes?*

c. *There are many accessor functions that provide programmatic access to the results contained in a GOHyperGResult instance. Read the help page for* HyperGResult-accessors *and explore the available accessor functions*

14.4.3   Generating an HTML report of test results

To make it easier for nontechnical users to review the results, the htmlReport function generates a HTML file that can be viewed in any Web browser.

```
> htmlReport(hgOver, file="ALL_hgo.html")
```

Exercise 14.8
Use the browseURL *function to view the HTML report produced by* htmlReport.

14.4.4   Results in detail

You can explore the relationships among significant GO terms using the termGraphs function. This function returns a list where each element is a subgraph of the GO DAG. The subgraphs consist of nodes that are connected in the DAG, and where all nodes are significant, by the Hypergeometric test.

```
> sigSub = termGraphs(hgOver)
```

Exercise 14.9
How many connected graphs did termGraphs *return? How many terms are in each graph?*

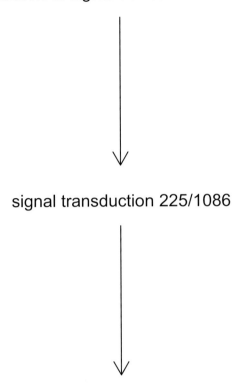

regulation of signal transduction 51/182

signal transduction 225/1086

cell communication 241/1170

Figure 14.1. The three terms shown above all came up as significant using the standard Hypergeometric test. Arrows in the plot point from child (more specific) to parent (more general) term. The values printed after the node name are: the number of interesting genes annotated at that node/the number of genes annotated at that node.

We show one of these subgraphs in Figure 14.1, where there are three significant terms. The first has 51 interesting genes out of 182, the second has those 51 plus another 174 interesting genes (total of 225), and the third has the 225 from the signal transduction node, plus an additional 16 (probably from other nodes that are not shown). When comparing the signal transduction to the cell communication node, we see that although there are 94 more genes annotated at the cell communication only 16 of them are in our list of interesting genes. And one must wonder whether that is substantial additional information, or whether we are merely seeing the effect for the `signal transduction` node again. In the next section, we use the conditional test to try to answer this question.

```
> plotGOTermGraph(sigSub[[1]], hgOver, max.nchar=100)
```

14.5 The conditional hypergeometric test

Figure 14.1 shows that there is considerable overlap in the genes annotated at *signal transduction* and *cell communication* and this leads to the following question. Is there additional evidence to mark the *cell communication* term significant beyond that provided by its significant child?

One way to answer this question is to perform a conditional Hypergeometric test that uses the relationships among the GO terms to adjust the results. We have implemented a method that conditions on all child terms which are themselves significant at a specified p-value cutoff. Given a subgraph of one of the three GO ontologies, we test the leaves of the graph, that is, those terms with no child terms. Before testing the terms whose children have already been tested, we remove all genes annotated at significant children from the parent's gene list. This continues until all terms have been tested. The procedure is described in Falcon and Gentleman (2007) and is similar to the approach described in Alexa et al. (2006), but was developed independently.

To perform a conditional test, we need to create a new *GOHyperGParams* instance with the `conditional` slot set to `TRUE`. This provides an opportunity to demonstrate the convenience of the parameter object design. Instead of having to call `hyperGTest` with an almost identical argument list, which could be error prone, we can simply make a copy of the parameter object and modify the relevant parts.

```
> paramsCond = params
> conditional(paramsCond) = TRUE
```

A similar approach works to create a parameter object for testing a different GO ontology or to create an object for testing under- rather than overrepresentation.

```
> hgCond = hyperGTest(paramsCond)
```

```
> hgCond
Gene to GO BP Conditional test for over-representation
1841 GO BP ids tested (4 have p < 0.001)
Selected gene set size: 713
```

```
      Gene universe size: 4229
      Annotation package: hgu95av2
```

Exercise 14.10

    a. Summarize the result of the conditional test using `summary`.

    b. List the GO terms that are marked significant by the standard Hypergeometric test and not by the conditional test.

Now let's look at the conditional results for the three terms shown in Figure 14.1.

```
> terms = nodes(sigSub[[1]])
> df = summary(hgCond, pvalue=0.5)[ , c("Term", "Pvalue")]
> df$Pvalue = round(df$Pvalue, 3)
> df$Term = sapply(df$Term, function(x) {
        if (nchar(x) <= 20) x
        else paste(substr(x, 1, 20), "...", sep="")
  })
> df[terms, ]
                            Term Pvalue
GO:0007154          cell communication  0.000
GO:0007165         signal transduction  0.005
GO:0009966 regulation of signal...  0.000
```

In the conditional test, the middle term, *signal transduction* is no longer significant at the 0.001 level suggesting that there is not enough evidence beyond that provided by the annotation from *regulation of signal* to claim significance of the more general term. We also observe that the parent term, *cell communication*, remains significant in the conditional analysis. This is because the implementation only conditions on a node's children, not all of its offspring.

14.6 Other collections of gene sets

Instead of testing for overrepresentation of GO terms, one can test other gene set collections such as chromosome band annotation, KEGG, and PFAM. The **Category** code was designed to make it easy for developers to add new category databases. We do note that the conditional testing is not always applicable, and some care should be taken if using it on gene sets other than those for GO.

In the exercises below, the main difference between different gene sets is the class used to represent the necessary parameters. They are: *ChrMapHyperGParams* for chromosome band testing, *KEGGHyperGParams* for KEGG pathways, and *PFAMHyperGParams* for PFAM domains.

14.6.1   Chromosome bands

The chromosome bands form a tree that can be used similarly to the GO graph for performing a test for overrepresentation. Because of the similar hierarchical structure, the same conditional analysis described above can be applied to a chromosome band analysis.

Exercise 14.11
Create a ChrMapHyperGParams object and perform a test for over representation of chromosome bands using a standard and a conditional test scheme.

14.6.2   KEGG

The Kyoto Encyclopedia of Genes and Genomes (KEGG), Kanehisa et al. (2006), provides, among other resources, a mapping of genes to pathways. As with GO, this information is included in most Bioconductor annotation packages.

Exercise 14.12
Determine if any KEGG pathways are overrepresented in the gene list. Are there any pathways that are underrepresented? Hint: create a KEGGHyperGParams instance.

14.6.3   PFAM

PFAM (Finn et al., 2006) is a database providing information about protein binding domains. Most Bioconductor annotation data packages contain a `PFAM` map linking genes to PFAM identifiers. The `hyperGTest` function can be used to assess overrepresentation of one or more protein binding domains.

Exercise 14.13
Repeat the test procedure using the PFAM data to define the gene set collection.

15

Solutions to Exercises

2 R and Bioconductor Introduction

Exercise 2.1

a > `apropos("plot")`

```
[1] ".__C__recordedplot"
[2] ".__M__KEGGmnplot:annotate"
[3] ".__M__MAplot:affy"
[4] "..."
```

b > `help.search("mann-whitney")`

c > `library("Biobase")`
  > `openVignette("Biobase")`

Exercise 2.2

`sessionInfo` prints version information about R and all loaded packages. This is helpful when posting on one of the R or Bioconductor mailing lists in order to provide detailed information about the software you are using.

> `sessionInfo()`

```
R version 2.7.0 Under development (unstable) (2007-10-16
 r43183)
i686-pc-linux-gnu
```

F. Hahne et al., *Bioconductor Case Studies*, DOI: 10.1007/978-0-387-77240-0_15,
© Springer Science+Business Media, LLC 2008

```
locale:
C

attached base packages:
[1] tools      stats      graphics  grDevices datasets
[6] utils      methods    base

other attached packages:
 [1] geneplotter_1.17.4    lattice_0.17-4
 [3] annotate_1.17.4       xtable_1.5-2
 [5] AnnotationDbi_1.1.9   RSQLite_0.6-4
 [7] DBI_0.2-4             hgu95av2cdf_2.0.0
 [9] hgu95av2probe_2.0.0   matchprobes_1.11.0
[11] CLL_1.2.4             affy_1.17.3
[13] preprocessCore_1.1.5  affyio_1.7.9
[15] RColorBrewer_1.0-2    GO_2.0.1
[17] class_7.2-39          hgu95av2_2.0.1
[19] BiocCaseStudies_1.1.2 Biobase_1.17.8
[21] weaver_1.5.0          codetools_0.1-3
[23] digest_0.3.1

loaded via a namespace (and not attached):
[1] KernSmooth_2.22-21 grid_2.7.0
```

Exercise 2.3

a
```
> x = c(0.1, 1.1, 2.5, 10)
> y = 1:100
> z = y < 10
> pets = c(Rex="dog", Garfield="cat", Tweety="bird")
```

b Arithmetic expressions in R are vectorized. The operations are performed element by element. If two vectors of unequal length are used in the same expression, R recycles the shorter of the two vectors.

```
> 2 * x + c(1,2)
[1]  1.2  4.2  6.0 22.0
```

c Index vectors can be of type *logical*, *integer*, and *character* (for the special case of named vectors).

```
> ##logical
> y[z]

[1] 1 2 3 4 5 6 7 8 9

> ## integer
> y[1:4]

[1] 1 2 3 4

> y[-(1:95)]

[1]  96  97  98  99 100

> ## character
> pets["Garfield"]

Garfield
   "cat"
```

Matrices and arrays can be indexed similarly to vectors. Each dimension is separated by a comma in the square brackets.

```
> m = matrix(1:12, ncol=4)
> m[1,3]

[1] 7
```

d List items are selected using the $ operator or the [[operator. The latter accepts all three types of index vectors; the former always interprets its right-hand argument literally as a name. Note that [returns a list even if only one element is selected. You can use the [[operator to get to the content of a single list element. Lists are created using the list function.

```
> l = list(name="Paul", sex=factor("male"), age=35)
> l$name

[1] "Paul"

> l[[3]]

[1] 35
```

e A *matrix* is a rectangular table of elements of equal type. In a *data.frame*, each column may have different type. R matrices and arrays are implemented as vectors with a dimension attribute, and data frames as a list of vectors that are all enforced to have the same length, but may be of different type.

Exercise 2.4

```
> ppc = function(x) paste("^", x, sep="")
```

Exercise 2.5

```
> myFindMap = function(mapEnv, which) {
    myg = ppc(which)
    a1 = eapply(mapEnv, function(x)
        grep(myg, x, value=TRUE))
    unlist(a1)
}
```

Exercise 2.6

```
a > theEnv = new.env(hash=TRUE)
  > theEnv$locations = myFindMap(hgu95av2MAP, 18)
```

```
b > theEnv$strip = function(x) gsub("18", "", x)
```

```
c > myExtract = function(env)  env$strip(env$locations)
  > myExtract(theEnv)[1:5]
    420_at 36469_at   808_at   862_at 35817_at
   "p11.2"    "q12" "q21.2"  "q21.3"    "q23"
```

Exercise 2.7

```
a > class(pData)
  [1] "data.frame"
```

```
b > names(pData)
  [1] "gender" "type"    "score"
```

```
c > sapply(pData, class)
     gender      type     score
   "factor"  "factor" "numeric"
```

```
d > pData[c(15, 20), c("gender", "type")]

    gender type
  O Female Case
  T Female Case

  > pData[pData$score > 0.8,]

    gender    type score
  E Female    Case  0.93
  G   Male    Case  0.96
  X   Male Control  0.98
  Y Female    Case  0.94
```

Exercise 2.8

```
> plot(x=x, y=y, log="xy",
       xlab="gene expression sample #1",
       ylab="gene expression sample #3",
       main="scatterplot of expression intensities",
       pch=20)
> abline(a=0, b=1)
```

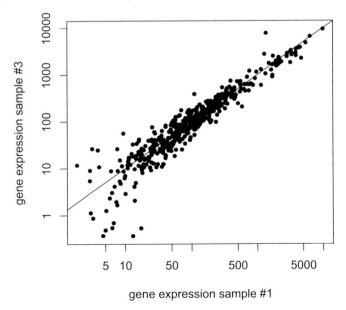

Figure 15.1. Scatterplot of expression intensities for two samples.

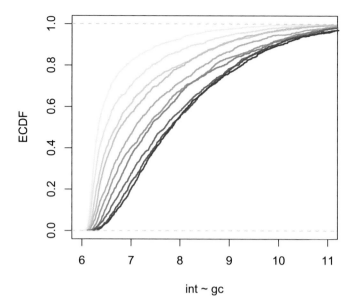

Figure 15.2. ECDF plot of distributions of $\log_2$-intensities from the CLL dataset grouped by GC-content.

Exercise 2.9

```
> multiecdf(int ~ gc, data=subset(probedata, gc %in% gcUse),
      xlim=c(6, 11), col=colorfunction(12)[-(1:2)],
      lwd=2, main="", ylab="ECDF")
```

3   Processing Affymetrix Expression Data

Exercise 3.1

```
> dataPLMx = fitPLM(CLLB)
> boxplot(dataPLM, main="NUSE", ylim = c(0.95, 1.3),
      outline = FALSE, col="lightblue", las=3,
      whisklty=0, staplelty=0)
> Mbox(dataPLM, main="RLE", ylim = c(-0.4, 0.4),
      outline = FALSE, col="mistyrose", las=3,
      whisklty=0, staplelty=0)
```

Exercise 3.2

There are lots of ways to do that, some of them are listed below.

```
> dim(e)[1]
[1] 12625
> nrow(e)
[1] 12625
> dim(exprs(CLLrma))[1]
[1] 12625
> nrow(CLLrma)
Features
   12625
> length(featureNames(CLLrma))
[1] 12625
```

Exercise 3.3

```
> par(mfrow=c(1,2))
> myPlot = function(...){
      plot(y = CLLtt$dm, pch = ".", ylim = c(-2,2),
          ylab = "log-ratio", ...)
      abline(h=0, col="blue")
  }
> myPlot(x = a, xlab="average intensity")
> myPlot(x = rank(a), xlab="rank of average intensity")
```

Exercise 3.4

Plot the two. Perhaps also use an ROC curve?

```
> plot(CLLtt$statistic, CLLeb$t[,2], pch=".")
```

Exercise 3.5

```
> plot(CLLtt$dm, -log10(CLLeb$p.value[,2]), pch=".",
      xlab="log-ratio", ylab=expression(log[10]~p))
> abline(h=2)
```

Exercise 3.6

```
> plot(CLLtt$dm, lod, pch=".", xlab="log-ratio",
       ylab=expression(log[10]~p))
> o1 = order(abs(CLLtt$dm), decreasing=TRUE)[1:25]
> points(CLLtt$dm[o1], lod[o1], pch=18, col="blue")
```

Exercise 3.7

```
> sum(CLLtt$p.value<=0.01)
[1] 243
> sum(CLLeb$p.value[,2]<=0.01)
[1] 261
```

Exercise 3.8

The values, transformed to a $\log_2$ scale, can be plotted using the code below.

```
> smoothScatter(log2(mms[,1]), log2(pms[,1]),
       xlab=expression(log[2] * "MM values"),
       ylab=expression(log[2] * "PM values"), asp=1)
> abline(a=0, b=1, col="red")
```

Let us look at their relative size.

```
> table(sign(pms-mms))
     -1       0       1
1414590   31828 2993182
```

In a large number of cases, the MM value is larger than the PM value. The simple story of MM measuring nonspecific hybridization and PM the sum of nonspecific and specific hybridization is hard to hold.

Exercise 3.9

The two histograms look very different. And we can confirm that, as suggested by the scatterplot in Figure 3.8, the intensities of the MM probes strongly correlate to those of the PM probes. The histogram for low values is quite skewed, whereas that corresponding to larger PM values is more symmetric.

```
> grouping = cut(log2(pms)[,1], breaks=c(-Inf, log2(2000),
       Inf), labels=c("Low", "High"))
```

```
> multidensity(log2(mms)[,1] ~ grouping, main="", xlab="",
      col=c("red", "blue"), lwd=2)
> legend("topright", levels(grouping), lty=1, lwd=2,
      col=c("red", "blue"))
```

Exercise 3.10

First, we create a subset sel of 500 randomly selected PM probes; this is enough to sample the background-correction transformation and reduces the file size of the plots.

```
> sel = sample(unlist(indexProbes(CLLB, "pm")), 500)
> sel = sel[order(exprs(CLLB)[sel, 1])]
```

Then we create the vectors yo, yr, and yv with the original, RMA background-corrected, and VSN background-corrected intensities for the first array,

```
> yo = exprs(CLLB)[sel,1]
> yr = exprs(bgrma)[sel,1]
> yv = exprs(bgvsn)[sel,1]
```

and plot them. The result is shown in Figure 3.10.

```
> par(mfrow=c(1,3))
> plot(yo, yr, xlab="Original", ylab="RMA", log="x",
      type="l", asp=1)
> plot(yo, yv, xlab="Original", ylab="VSN", log="x",
      type="l", asp=1)
> plot(yr, yv, xlab="RMA", ylab="VSN", type="l", asp=1)
```

Exercise 3.11

We need to pay attention to the fact that the nonspecific filtering selected different sets of probe sets. In inboth, we determine those that are in common.

```
> inboth = intersect(featureNames(CLLvsnf),
                    featureNames(CLLf))
```

```
> plot(CLLtt[inboth, "statistic"],
      CLLvsntt[inboth, "statistic"],
      pch=".", xlab="RMA", ylab="VSN", asp=1)
```

The scatterplot is shown in Figure Chapter 3.11.

Exercise 3.12

We can use the `matplot` function to do this. You should probably either transform the data to the log scale, or use log-scaling in the plot, as we have done. PMs are plotted using a P and MMs using a M. It is worth noting that for many of the probes, there is no clear separation between the MM values and the PM values (e.g. probe 1); for others the MM values seem to be higher(!) than the PM values (e.g. probe 3), and for others the PM values are larger than the MM values.

```
> colors = brewer.pal(8, "Dark2")
> Index = indices[["189_s_at"]][seq(along=colors)]
> matplot(t(pms[Index, 1:12]), pch="P", log="y", type="b",
      lty=1, main="189_s_at", xlab="samples",
      ylab=expression(log[2]~Intensity),
      ylim=c(50,2000), col=colors)
> matplot(t(mms[Index, 1:12]), pch="M", log="y", type="b",
      lty=3, add=TRUE, col=colors)
```

The result is shown in Figure Chapter 3.12.

Exercise 3.13

We can compute the percentage, for each array, by first creating a logical matrix where TRUE corresponds to a negative value and FALSE corresponds to a nonnegative value. Then the column sums of that matrix are the proportions, and if we multiply by 100 we get percentages.

```
> colMeans(newsummary<0)*100
 [1] 20.2 19.6 19.4 18.3 21.0 22.6 21.7 19.6 21.7 21.1
[11] 18.9 18.7 20.6 23.1 19.6 21.0 18.6 21.6 21.4 19.6
[21] 19.7 19.8
```

4   Two-Color Arrays

Exercise 4.1

There are 18 files with the extension `.gpr`. They contain the output of the image analysis, that is, the quantified red and green intensities for each feature on the arrays. The 18 files correspond to the 18 arrays. A description of what was hybridized to these arrays is in the file `samplesInfo.txt`.

Exercise 4.2

a By subtracting the background estimate, many of the resulting values are negative, and the application of the logarithm leads to `NA` values. The scatterplot then also makes the imbalances between the red and the green color channels more apparent; this is seen in the more pronounced curved shape of the distribution.

b `plotMA` plots each data point as a dot, whereas `smoothScatter` plots the density of the data points with a smooth false color map. In data-dense regions, the latter may be more informative. When the number of data points is large, `smoothScatter` also uses less time to display and the plot is more compact to store in the vector graphics file format produced by R's PDF device.

5 Fold-Changes, Log-Ratios, Background Correction, Shrinkage Estimation, and Variance Stabilization

Exercise 5.1

a
```
> r1 = rnorm(10000, mean=2000, sd=50)
> g1 = rnorm(10000, mean=1000, sd=50)
> hist(r1/g1, breaks=33, col="azure")
```

b
```
> r2 = rnorm(10000, mean=200, sd=50)
> g2 = rnorm(10000, mean=100, sd=50)
> ratio = r2/g2
> ratio = ratio[(ratio>0)&(ratio<6)]
> hist(ratio, breaks=33, col="azure",
        main="Histogram of r2/g2", xlab="r2/g2",
        sub="restricted to [0,6]")
```

c
```
> hist(log2(r1/g1), breaks=33, col="azure")
```

```
> hist(log2(r2/g2), breaks=33, col="azure")
```

Exercise 5.2

For the kidney data:

```
> library("geneplotter")
> pcol = c("green3", "red1")
> plty = 1:2
> multidensity(exprs(kidney), xlim=c(-200, 1000),
      main = "kidney", xlab="Intensity",
      lty = plty, col = pcol, lwd = 2)
> legend("topright", c("green", "red"),
      lty = plty, col = pcol, lwd = 2)
```

For the CCl$_4$ data:

```
> multidensity(cbind(assayData(CCl4s)$G[,1],
      assayData(CCl4s)$R[,1]), xlim=c(0, 200),
      main = expression(CCl[4]), lwd=2, xlab="Intensity",
      col = pcol, lty = plty)
```

Exercise 5.3

```
a > px = seq(-100, 500, length=50)
  > f  = function(x, b) log2(x+b)
  > h  = function(x, a) log2((x+sqrt(x^2+a^2))/2)
  > matplot(px, y=cbind(h(px, a=50), f(px, b=50)),
      type="l", lty=1:2, xlab="x", ylab="f, h")
```

b Repeat the plot command above, with

```
  > px = seq(0, 1e8, length=50)
```

c This question is very nicely explored in a paper by Rocke and Durbin (2003).

Exercise 5.4

```
> axl = c(30, 300)
> plot(assayData(CCl4s)$R[,1],
      assayData(CCl4s)$G[,1],
      xlim=axl, ylim=axl, pch=".", col="grey",
      asp=1)
> abline(a=0, b=1, col="blue", lty=2, lwd=3)
> abline(a=18, b=1.2, col="red", lty=3, lwd=3)
```

The comparison of the dashed line with the data shows that there are systematic dye-related differences between the two channels. The dotted line shows that they can be fitted by an affine transformation, that is, a shift and a scaling of the data from one dye to adjust them to the other.

6   Easy Differential Expression

Exercise 6.1

The histogram indicates that the nonspecific filtering step did not remove a large number of probe sets that would have been detected as differentially expressed. Hence, we have managed to enrich the filtered set of 8812 probe sets in `ALLsfilt` for those that are potentially differentially expressed, and to deplete it from uninformative "background" probe sets. We have reduced the size of multiple testing adjustments without a major loss in sensitivity.

7   Differential Expression

Exercise 7.1

a There is a weak relationship but it is not dominant. We may safely proceed with the nonspecific filtering based on variablility.

b It switches between plotting the x-axis (means) on the original scale (`FALSE`) or on the rank scale (`TRUE`). The latter distributes the data more evenly along the x-axis and allows a better visual assessment of the standard deviation as a function of the mean.

Exercise 7.2

The number of probe sets with p-value less than 0.05 and mean $\log_2$ fold-change larger than 0.5 is

```
> sum(tt$p.value<0.05 & abs(tt$dm)>0.5)
[1] 224
```

This choice of thresholds is of course arbitrary.

Exercise 7.3

```
> mtyp = ALLset1$mol.biol
> sel = rep(1:2, each=rev(table(mtyp)))
> plot(exprs(ALLset1)[j, order(mtyp)], pch=c(1,15)[sel],
```

```
    col=c("black", "red")[sel],
    main=featureNames(ALLset1)[j],
    ylab=expression(log[2]~expression~level))
> legend("bottomleft", col=c("black", "red"),
    pch=c(1,15), levels(mtyp), bty="n")
```

Exercise 7.4

The curve for a bad discriminator would be close to the diagonal because the classification would be almost random. The curve for a perfect discriminator shows both high sensitivity and high specificity over the whole plot, that is, a rectangle from [0,1] to [1,1].

Exercise 7.5

The identification of differentially expressed genes by area under the ROC curve is not so much affected by the sample size as the t-statistic is. For the t-test the number of differentially expressed genes increases constantly with the sample size. For the ROC curves this number stabilizes with a sufficient sample size.

8   Annotation and Metadata

Exercise 8.1

```
> hist(rt$statistic, breaks=100, col="skyblue")
```

```
> hist(rt$p.value, breaks=100, col="mistyrose")
```

Exercise 8.2

```
> sel = order(rt$p.value)[1:400]
> ALLsub = ALLfilt_af4bcr[sel,]
```

Exercise 8.3

First, we map from the Affymetrix identifiers to EntrezGene IDs.

```
> EG    = as.character(hgu95av2ENTREZID[featureNames(ALL)])
> EGsub = as.character(hgu95av2ENTREZID[featureNames(ALLsub)])
```

Then, we find the multiplicity by a frequently used and efficient idiom of the R language. The two calls to the function `table` work as follows. The inner one counts for each EntrezGene ID the number of probe sets that are mapped to it. The outer one tabulates how often each count is seen, one, two, three, ... times.

```
> table(table(EG))
   1    2    3    4    5    6    7    8    9
6897 1555  506  101   23   12    8    5    1
> table(table(EGsub))
   1
 400
```

There are 6897 instances of EntrezGene IDs that are matched by exactly one probe set in `ALL`, whereas 1555 EntrezGene IDs are matched by two probe sets. That the probe sets in `ALLsub` all map to a unique EntrezGene ID is no coincidence. This has been achieved by our call to the `nsFilter` function above (type `? nsFilter` to find out more about this).

Exercise 8.4

```
> syms = as.character(hgu95av2SYMBOL[featureNames(ALLsub)])
> whFeat = names(which(syms =="CD44"))
> ordSamp = order(ALLsub$mol.biol)
> CD44 = ALLsub[whFeat, ordSamp]
> plot(as.vector(exprs(CD44)), main=whFeat,
        col=c("sienna", "tomato")[CD44$mol.biol],
        pch=c(15, 16)[CD44$mol.biol], ylab="expression")
```

Exercise 8.5

First, we create the *data.frame* z that contains the mapping between probe sets and chromosome identifiers; then we use the function `table` to produce the table of frequencies.

```
> z = toTable(hgu95av2CHR[featureNames(ALLsub)])
> chrtab = table(z$chromosome)
> chrtab
 1 10 11 12 13 14 15 16 17 18 19  2 20 21 22  3  4  5  6  7
40 21 23 20  9 20  7 12 16  6 14 26  9  7 13 21 13 13 42 21
 8  9  X  Y
13 20 14  1
```

To plot the frequencies entries in the numeric order of the chromosomes, we need one extra step constructing `chridx`, as in the code below.

```
> chridx = sub("X", "23", names(chrtab))
> chridx = sub("Y", "24", chridx)
> barplot(chrtab[order(as.integer(chridx))])
```

Exercise 8.6

First, we compute a list probeSetsPerGene that contains, for each Entrez-Gene ID, the list of probe sets that are mapped to it.

```
> probeSetsPerGene = split(names(EG), EG)
> j = probeSetsPerGene$"7013"
> j
[1] "1329_s_at"  "1342_g_at"  "1361_at"    "32255_i_at"
[5] "32256_r_at" "32257_f_at" "32258_r_at"
```

Then we plot the data for the first and seventh probe set of Entrez Gene ID 7013.

```
> plot(t(exprs(ALL_af4bcr)[j[c(1,7)], ]), asp=1, pch=16,
      col=ifelse(ALL_af4bcr$mol.biol=="ALL1/AF4", "black",
      "grey"))
```

We can also consider a heatmap.

```
> library("lattice")
> mat = exprs(ALL_af4bcr)[j,]
> mat = mat - rowMedians(mat)
> ro = order.dendrogram(as.dendrogram(hclust(dist(mat))))
> co = order.dendrogram(as.dendrogram(hclust(dist(t(mat)))))
> at = seq(-1, 1, length=21) * max(abs(mat))
> lp = levelplot(t(mat[ro, co]),
    aspect = "fill", at = at,
    scales = list(x = list(rot = 90)),
    colorkey = list(space = "left"))
> print(lp)
```

What is the effect of the median centering? What does the heatmap look like if you do not do the centering?

Exercise 8.7

```
> ps_chr = toTable(hgu95av2CHR)
> ps_eg  = toTable(hgu95av2ENTREZID)
> chr = merge(ps_chr, ps_eg)
```

```
> chr = unique(chr[, colnames(chr)!="probe_id"])
> head(chr)
  chromosome gene_id
1         16    5595
2          1    7075
3         10    1557
4         11     643
6          5    1843
7         11    4319
```

We see that in `chr` some EntrezGene IDs are mapped to multiple chromosomes (you might want to investigate which ones):

```
> table(table(chr$gene_id))

   1    2
9030   12
```

Here, for simplicity, we just remove conflicting mappings.

```
> chr = chr[!duplicated(chr$gene_id), ]
```

Exercise 8.8

```
> isdiff = chr$gene_id %in% EGsub
> tab = table(isdiff, chr$chromosome)
> tab
isdiff    1   10   11   12   13   14   15   16   17   18   19    2   20
  FALSE 907  309  506  478  151  270  252  366  515  122  548  547  221
  TRUE   40   21   23   20    9   20    7   12   16    6   14   26    9

isdiff   21   22    3    4    5    6    7    8    9    X    Y
  FALSE  94  249  466  331  388  498  411  298  313  377   25
  TRUE    7   13   21   13   13   42   21   13   20   14    0
> fisher.test(tab, simulate.p.value=TRUE)

    Fisher's Exact Test for Count Data with simulated
    p-value (based on 2000 replicates)

data:  tab
p-value = 0.02449
alternative hypothesis: two.sided
> chisq.test(tab)

    Pearson's Chi-squared test
```

```
data:  tab
X-squared = 40.6, df = 23, p-value = 0.01324
```

Exercise 8.9

```
> chrloc = toTable(hgu95av2CHRLOC[featureNames(ALLsub)])
> head(chrloc)
   probe_id start_location Chromosome
1 1039_s_at       61231991         14
2   1065_at      -27475410         13
3    106_at      -25098589          1
4   1113_at        6696744         20
5   1118_at       40715788          5
6   1135_at      120957186         10
```

A little complication arises because some genes, and hence some probe sets, have multiple (alternative) transcription start sites and therefore are annotated at multiple locations.

```
> table(table(chrloc$probe_id))
  1   2   3   4   5   6   9
343  34  14   4   1   1   1
```

We can collapse this table such that for each probe set we only record the strand, which is unique.

```
> strds = with(chrloc,
    unique(cbind(probe_id, sign(start_location))))
> table(strds[,2])
 -1   1
192 206
```

Exercise 8.10

We call the summary method, with $p = 0.001$.

```
> sum = summary(mfhyper, p=0.001)
> head(sum)
                    GOBPID   Pvalue OddsRatio ExpCount Count Size
GO:0006955 GO:0006955 2.93e-08      2.65   24.731    53  264
GO:0007154 GO:0007154 6.62e-07      1.72  111.101   154 1186
GO:0007165 GO:0007165 1.15e-06      1.72  102.951   144 1099
GO:0019882 GO:0019882 2.61e-06      6.07    3.466    14   37
GO:0002376 GO:0002376 3.02e-06      2.12   32.881    59  351
```

```
GO:0006687 GO:0006687 4.21e-06      58.92      0.656      6      7
                                                Term
GO:0006955                          immune response
GO:0007154                       cell communication
GO:0007165                      signal transduction
GO:0019882 antigen processing and presentation
GO:0002376                    immune system process
GO:0006687 glycosphingolipid metabolic process
```

In total, the table sum contains 25 categories. Several relate to the immune system and lymphocyte proliferation. This is not surprising given the role that B-cells play and the fact that the disease studied is a leukemia.

Exercise 8.11

For each GO identifier, an object of class *GOTerms* can be retrieved from the GOTERM annotation object that is supplied in the **GO.db** package. It contains various pieces of information about that category, as shown below.

```
> GOTERM[["GO:0032945"]]
GOID: GO:0032945
Term: negative regulation of mononuclear cell
    proliferation
Ontology: BP
Definition: Any process that stops, prevents or
    reduces the frequency, rate or extent of
    mononuclear cell proliferation.
Synonym: negative regulation of PBMC proliferation
Synonym: negative regulation of peripheral blood
    mononuclear cell proliferation
```


Exercise 8.12

```
> utr = getSequence(id=EGsub, seqType="3utr",
      mart=ensembl, type="entrezgene")
> utr[1,]
            1
3utr        "CTTCGTTTTTGATTGTGTTGGTGTC..."
entrezgene "10950"
            2
3utr        "ATTATTCAGTGCCACAAATTGAAAG..."
entrezgene "11034"
            3
```

```
3utr          "CTCTCTGCTGAATATGGGGTTGGTG..."
entrezgene "  241"
           4
3utr          "ATCAGGAGGCATCACTGAGGCCAGG..."
entrezgene "10410"
           5
3utr          "AGGAACAATTTAGTTTTAAGGACTT..."
entrezgene " 2322"
```

Exercise 8.13

```
> domains = getBM(attributes=c("entrezgene", "pfam",
      "prosite", "interpro"), filters="entrezgene",
      value=EGsub, mart=ensembl)
> interpro = split(domains$interpro, domains$entrezgene)
> interpro[1]
$`25`
 [1] "IPR000719" "IPR008266" "IPR000980" "IPR001245"
 [5] "IPR001452" "IPR001720" "IPR011511" "IPR015015"
 [9] "IPR000719" "IPR008266" "IPR000980" "IPR001245"
[13] "IPR001452" "IPR001720" "IPR011511" "IPR015015"
[17] "IPR000719" "IPR008266" "IPR000980" "IPR001245"
[21] "IPR001452" "IPR001720" "IPR011511" "IPR015015"
[25] "IPR000719" "IPR008266" "IPR000980" "IPR001245"
[29] "IPR001452" "IPR001720" "IPR011511" "IPR015015"
[33] "IPR000719" "IPR008266" "IPR000980" "IPR001245"
[37] "IPR001452" "IPR001720" "IPR011511" "IPR015015"
[41] "IPR000719" "IPR008266" "IPR000980" "IPR001245"
[45] "IPR001452" "IPR001720" "IPR011511" "IPR015015"
[49] "IPR000719" "IPR008266" "IPR000980" "IPR001245"
[53] "IPR001452" "IPR001720" "IPR011511" "IPR015015"
[57] "IPR000719" "IPR008266" "IPR000980" "IPR001245"
[61] "IPR001452" "IPR001720" "IPR011511" "IPR015015"
[65] "IPR000719" "IPR008266" "IPR000980" "IPR001245"
[69] "IPR001452" "IPR001720" "IPR011511" "IPR015015"
[73] "IPR000719" "IPR008266" "IPR000980" "IPR001245"
[77] "IPR001452" "IPR001720" "IPR011511" "IPR015015"
[81] "IPR000719" "IPR008266" "IPR000980" "IPR001245"
[85] "IPR001452" "IPR001720" "IPR011511" "IPR015015"
[89] "IPR000719" "IPR008266" "IPR000980" "IPR001245"
[93] "IPR001452" "IPR001720" "IPR011511" "IPR015015"
```

Exercise 8.14

We can use the same type of query as for finding terms that contain the word chromosome. The % wild card matches zero or more arbitrary characters, hence we are looking for all terms that contain the words *transcription factor* at their beginning, in the middle, or in the end.

```
> query = paste("select term from go_term where term",
      "like '%transcription factor%'")
> tf = dbGetQuery(GO_dbconn(), query)
> nrow(tf)
[1] 35
> head(tf)
                                                    term
1    RNA polymerase I transcription factor complex
2                transcription factor TFIIIB complex
3                transcription factor TFIIIC complex
4                      transcription factor activity
5   RNA polymerase I transcription factor activity
6 RNA polymerase II transcription factor activity
```

9   Supervised Machine Learning

Exercise 9.1

```
> table(ALL_bcrneg$mol.biol)
BCR/ABL     NEG
     37      42
```

Exercise 9.2

```
> class(ALLfilt_bcrneg)
[1] "ExpressionSet"
attr(,"package")
[1] "Biobase"
```

Exercise 9.3

The distances available include Kullback-Leibler distance, mutual information distance, Euclidean distance, Manhattan distance, and correlation distance (using Pearson, Spearman, or Kendall's tau). See the `dist` function and the `daisy` function in the **cluster** package for other distances.

Exercise 9.4

The diagonal band of blue squares is probably the most prominent feature, but it is merely indicating that each sample is distance zero from itself. After that you might notice that there are some bluish colored blocks along the diagonal, the most prominent one being in the top-left corner. The dendrogram also suggests that those samples are similar to each other, and some distance from the others.

Exercise 9.5

Because the **bioDist** package is loaded we can simply call the `spearman.dist` function. All other steps are essentially the same, as before.

```
> spD = spearman.dist(ALLfilt_bcrneg)
> spD@Size
[1] 79
> spM = as.matrix(spD)
```

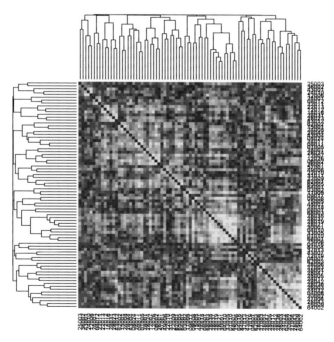

Figure 15.3. A heatmap of the between-sample distances, for the same data as in Figure 9.1, but now using Spearman's correlation instead of the Euclidean distance.

```
> heatmap(spM, sym=TRUE, col=hmcol,
      distfun=function(x) as.dist(x))
```

In this heatmap, the samples seem to be farther from each other (darker red colors predominate), but there are a small number that are quite close, as evidenced by the light blue rectangle in the middle of the heatmap.

Exercise 9.6

```
> cD = MIdist(ALLfilt_bcrneg)
> cM = as.matrix(cD)
> closest.top("03002", cM, 1)
[1] "09017"
```

Exercise 9.7

We use the MLearn interface to the machine learning code. We make use of the MLearn interface to the different machine learning tools, provided by MLearn.

The function, confuMat, can be used to compute the confusion matrix, and from that we can estimate the error rates.

```
> kans = MLearn( mol.biol ~ ., data=ALLfilt_bcrneg,
      knnI(k=1,l=0), TrainInd)
> confuMat(kans)
         predicted
given      BCR/ABL NEG
  BCR/ABL       14   3
  NEG            8  14
> dldans = MLearn( mol.biol ~ ., ALLfilt_bcrneg, dldaI,
      TrainInd)
> confuMat(dldans)
         predicted
given      BCR/ABL NEG
  BCR/ABL       12   5
  NEG            8  14
> ldaans = MLearn( mol.biol ~ ., ALLfilt_bcrneg, ldaI,
      TrainInd)
> confuMat(ldaans)
         predicted
given      BCR/ABL NEG
  BCR/ABL       12   5
  NEG            6  16
```

Exercise 9.8

    a Ties are broken at random. This suggests that it might not be all that helpful to select a value of k that is even, as different users would potentially classify samples differently, given the same data.

    b This is difficult with the current implementation. You would essentially need to do the nearest neighbor finding directly from the distance matrix, and this will be somewhat slow. The `closest.top` function, from the **bioDist** package could be used.

    c The `knn` function has a parameter `prob` that if set to `TRUE` will cause the proportion of votes for the winning class to be returned. This could be used. Also, the parameter `l` can be used; in that case doubt is encoded as NA. The concept of outlier is more difficult, but could potentially be handled in a preprocessing step. Any object that is a long way from all other objects could be identified as an outlier and removed. This does not help with pairs of outliers, or triples.

Exercise 9.9

We repeat the steps taken above, but use all of the data.

```
> alltt = rowttests(ALLfilt_bcrneg, "mol.biol")
> ordall = order(abs(alltt$statistic), decreasing=TRUE)
> fNall = featureNames(ALLfilt_bcrneg)[ordall[1:50]]
> intersect(fNall, fNtt)
 [1] "1635_at"    "1674_at"    "40504_at"   "37015_at"
 [5] "40202_at"   "32434_at"   "39837_s_at" "37403_at"
 [9] "40480_s_at" "41815_at"   "33774_at"   "36591_at"
[13] "34472_at"   "37014_at"   "31786_at"   "39329_at"
[17] "32542_at"   "33362_at"   "33440_at"   "40196_at"
[21] "40051_at"   "38032_at"   "40795_at"   "40516_at"
[25] "32134_at"   "40132_g_at" "671_at"     "35912_at"
[29] "36617_at"   "38994_at"
```

Exercise 9.10

We simply redo the calls with

```
>   dldtt = MLearn( mol.biol ~ ., BNf, dldaI, TrainInd)
>   confuMat(dldtt)
         predicted
given     BCR/ABL NEG
  BCR/ABL      10   7
  NEG           5  17
```

```
> ldatt  = MLearn( mol.biol ~ ., BNf, ldaI, TrainInd)
> confuMat(ldatt)
         predicted
given      BCR/ABL NEG
  BCR/ABL       12   5
  NEG            8  14
```

In all cases the error rates are lower, which is nice.

Exercise 9.11

Each sample is left out, in turn, and because $k = 1$ the class of that sample is determined by its nearest neighbor in the remaining $n - 1$ samples. For larger values of k, then more nearest neighbors would be used in the prediction. The confusion matrix is produced by the confuMat function, and it can be used to estimate either the overall error rate, or the class conditional error rates.

```
> knnCM = confuMat(knnXval1)
> knnCM
         predicted
given      BCR/ABL NEG
  BCR/ABL       31   6
  NEG           16  26
> #overall error rate
> (knnCM[1,2] + knnCM[2,1])/sum(knnCM)
[1] 0.278
> #class conditional error rates
> knnCM[1,2]/sum(knnCM[1,])
[1] 0.162
> knnCM[2,1]/sum(knnCM[2,])
[1] 0.381
```

So it seems that it was harder to predict the NEG phenotype than the BCR/ABL phenotype.

Exercise 9.12

```
a > lk3f2 = MLearn(mol.biol~., data=BNx, knnI(k=1),
        xvalSpec("LOO", fsFun=fs.absT(5)))
  > confuMat(lk3f2)
  > table(unlist(fsHistory(lk3f2)))
```

The error rate seems to be bit higher when only five features are selected.

Exercise 9.13

This is quite an interesting problem. Basically, what you need to do is to
try out the KNN algorithm, for a variety of values of k, and see what value
of k gives the lowest error rate.

```
> knnXval2 = MLearn(mol.biol~., data=BNx, knn.cvI(k=2, l=0),
      trainInd=1:ncol(BNx))
> confuMat(knnXval2)


> knnXval3 = MLearn(mol.biol~., data=BNx, knn.cvI(k=3, l=0),
      trainInd=1:ncol(BNx))
> confuMat(knnXval3)


> knnXval5 = MLearn(mol.biol~., data=BNx, knn.cvI(k=5, l=0),
      trainInd=1:ncol(BNx))
> confuMat(knnXval5)
```

Exercise 9.14

We are only concerned with the errors for the test set because those for the
training set are known to be overly optimistic. Error rates can be computed
for either all predictions combined, or on a per class basis. It is often the
case that error rates can be quite different for different classes, so we also
compute the class conditional error rates. Note that the error rates are a
bit worse for model 2, which had a much smaller value of `mtry`.

```
> cf1 = confuMat(rf1)
> overallErrM1 = (cf1[2,1] + cf1[1,2])/sum(cf1)
> overallErrM1
[1] 0.128
> perClass1 = c(cf1[1,2], cf1[2,1])/rowSums(cf1)
> perClass1
BCR/ABL      NEG
 0.0588   0.1818
```

And now for model 2.

```
> cf2 = confuMat(rf2)
> overallErrM2 = (cf2[2,1] + cf2[1,2])/sum(cf2)
> overallErrM2
[1] 0.179
> perClass2 = c(cf2[1,2], cf2[2,1])/rowSums(cf2)
> perClass2
```

```
BCR/ABL      NEG
  0.176    0.182
```

For KNN we had the following error rates:

```
> cfKNN = confuMat(knnf)
> (cfKNN[1,2] + cfKNN[2,1])/sum(cfKNN)
[1] 0.179
> #class conditional error rates
> cfKNN[1,2]/sum(cfKNN[1,])
[1] 0.118
> cfKNN[2,1]/sum(cfKNN[2,])
[1] 0.227
```

And in this case, it seems that KNN has the lower overall error rate, 0.179 compared to 0.128 for model 1 and 0.179 for model 2.

Exercise 9.15

We can obtain the importance measure by calling the importance function.

```
> impvars = function(x, which="MeanDecreaseAccuracy", k=10) {
      v1 = order(importance(x)[,which], decreasing=TRUE)
      importance(x)[v1[1:k],]
  }
> ivm1 = impvars(rf1@RObject, k=20)
> ivm2 = impvars(rf2@RObject, k=20)
> intersect(row.names(ivm1) , row.names(ivm2))
[1] "X39837_s_at" "X40132_g_at" "X1467_at"
[4] "X36638_at"   "X41815_at"   "X1674_at"
[7] "X38385_at"
```

The other importance measure is called MeanDecreaseGini, and we leave that part of the problem to the reader.

Exercise 9.16

Reversing the role of the test and training sets is quite simple, we use model 2.

```
> rfRev = MLearn( mol.biol~., data=ALLfilt_bcrneg,
      randomForestI, TestInd, ntree=2000, mtry=10,
      importance=TRUE)
> rfRev
```

MLInterfaces classification output container
The call was:
MLearn(formula = mol.biol ~ ., data = ALLfilt_bcrneg, me
thod = randomForestI,
    trainInd = TestInd, ntree = 2000, mtry = 10, importa
    nce = TRUE)
Predicted outcome distribution for test set:

BCR/ABL   NEG
    20   20

and for the confusion matrix

```
> cfR = confuMat(rfRev)
> cfR
        predicted
given     BCR/ABL NEG
  BCR/ABL      17   3
  NEG           3  17
> overallErr = (cfR[2,1] + cfR[1,2])/sum(cfR)
> overallErr
[1] 0.15
> perClass = c(cfR[1,2], cfR[2,1])/rowSums(cfR)
> perClass
BCR/ABL    NEG
   0.15   0.15
```

We can see from the confusion matrix that the error rate observed is
roughly comparable to that obtained with the other split, as we expected,
inasmuch as the two sets were roughly the same size.

It is also quite simple to use the whole dataset to fit a random forest.

```
> rfAll = MLearn( mol.biol~., data=ALLfilt_bcrneg,
      randomForestI, 1:79, ntree=1000, mtry=10,
      importance=TRUE)
> rfAll@RObject
Call:
 randomForest(formula = formula, data = trdata, ntree =
 1000,      mtry = 10, importance = TRUE)
               Type of random forest: classification
                     Number of trees: 1000
No. of variables tried at each split: 10

       OOB estimate of  error rate: 21.5%
```

```
Confusion matrix:
        BCR/ABL NEG class.error
BCR/ABL      26  11       0.297
NEG           6  36       0.143
```

Exercise 9.17

Using KNN is quite straightforward. We demonstrate its use for $k = 1$; you might want to try other methods.

```
> knn1MV = knn(t(exprs(trainSet)), t(exprs(testSet)),
      trainSet$mol.biol)
> tab1 = table(knn1MV, testSet$mol.biol)
> tab1
knn1MV     ALL1/AF4 BCR/ABL NEG
   ALL1/AF4        5       0   0
   BCR/ABL         0      16  13
   NEG             0       2   8

> s3 = table(testSet$mol.biol)
```

Class conditional error rates are estimated by considering those with the correct classification (on the diagonal of the table produced above). For example, of the 18 BCR/ABL samples in the test set, 16 are correctly classified, so that the class conditional error rate is 0.11.

It is not so easy to handle unbalanced data. One can, in principle, find k nearest neighbors, and then compare the proportion of nearest neighbors to the class counts.

10   Unsupervised Machine Learning

Exercise 10.1

a The three-dimensional reduction is obtained by specifying the parameter k in the call to sammon.

```
> sam2 = sammon(manDist, k=3,  trace=FALSE)
```

b The R function is cmdscale.

```
> cmd1 = cmdscale(manDist)
```

c This is somewhat more involved and requires the use of a number of tools. Plotting the points that are computed via Sammon mapping yields a fairly good separation of the data into two groups, which we expect, because we essentially forced that to be true by our selection of the genes with large *t*-statistics.

```
> rtt = rowttests(ALLfilt_bcrneg, "mol.biol")
> ordtt = order(rtt$p.value)
> esTT = ALLfilt_bcrneg[ordtt[1:50],]
> dTT = dist(t(exprs(esTT)), method="manhattan")
> sTT = sammon(dTT, trace=FALSE)
```

Exercise 10.2

We use the maximum distance, but you could have chosen any other. We suggest examining some of the distance measures in the **bioDist** package.

```
> dsol = as.matrix(dist(gvals), method="maximum")
> silcheck(dsol, diss=TRUE)
[1] 3.000 0.112
> msscheck(dsol)
[1] 3.0000 0.0575
```

Exercise 10.3

We first use cutree as described above.

```
> hc13 = cutree(hc1, k=3)
> hc23 = cutree(hc2, k=3)
> hc33 = cutree(hc3, k=3)
> hc43 = cutree(hc4, k=3)
```

And now we need to compare the outputs. It is relatively easy to do that for the *hclust* objects, and unfortunately less so for comparisons with the *diana* object.

```
> table(hc13, hc33)
     hc33
hc13  1  2  3
   1 22  0 15
   2  2 18  0
   3  1  0 21
```

Exercise 10.4

The code to compute the cophenetic correlations, and to compute the correlation with the original distances is given below.

```
> cph2 = cophenetic(hc2)
> cor2 = cor(manDist, cph2)
> cor2
[1] 0.396
> cph3 = cophenetic(hc3)
> cor3 = cor(manDist, cph3)
> cor3
[1] 0.494
> cph4 = cophenetic(hc4)
> cor4 = cor(manDist, cph4)
> cor4
[1] 0.503

> stopifnot( cor2 == min(cor1, cor2, cor3, cor4) )
```

We see that for single-linkage clustering the cophenetic correlation is much lower than for the other three, suggesting that it is a relatively poor choice of hierarchical clustering method. The other three values are quite similar.

Exercise 10.5

The values returned can be examined by using names and by checking the manual page. One of the components returned is the cluster allocation, and we can use the table command to see if there are different allocations.

```
> names(km2)
[1] "cluster"  "centers"  "withinss" "size"
> table(km2$cluster, kmx$cluster)

     1   2
  1   0  62
  2  17   0
```

Exercise 10.6

There are 21 phenotypic variables available. Of these you can find out which are factors by simply using an apply-type function, as is shown in the code below. Notice that we are going to use either variables that are explicitly factors, or those that are implicitly factors (because a logical variable can take only two values it is really a factor). We also do a quick check to make sure that samples in the expression set are in the same order as the values in the clustering (this is harder with the `pam` output because the clustering vectors from it are not named, (at least not as of version 1.11.6). If the value returned by the second command below is not `TRUE` then all other computations are going to be incorrect and some corrective action is needed.

```
> sapply(pData(es2), function(x) is.factor(x) ||
      is.logical(x) )
            cod          diagnosis               sex
          FALSE              FALSE              TRUE
            age                 BT          remission
          FALSE               TRUE              TRUE
             CR            date.cr            t(4;11)
          FALSE              FALSE              TRUE
        t(9;22)        cyto.normal              citog
           TRUE               TRUE             FALSE
       mol.biol     fusion protein                mdr
           TRUE               TRUE              TRUE
          kinet                ccr            relapse
           TRUE               TRUE              TRUE
      transplant         f.u date     last seen
           TRUE              FALSE             FALSE
```

Then for each of those variables, such as say, `mdr`, we can form a two-way table and use any one of your favorite tests for association. In the example below, we use the χ^2 test, but there are many others that you could use.

```
> tt1 = table(es2$mdr, km2$cluster)
> test1 = chisq.test(tt1)
> test1$p.value
[1] 0.0935
```

You can then repeat this for each categorical variable and select the one with the best p-value as that variable that most closely aligns with the clustering.

Exercise 10.7

The answer is a little bit tricky, because the cluster labels are completely arbitrary. So we first create a two-way table, showing how the clusters align. We want to make this a bit interesting so we compute a three cluster k-means solution to compare with the three cluster PAM solution from above.

```
>  km3 = kmeans(gvals, centers=3, nstart=25)

>  table(km3$cluster, pam3$clustering)
      1  2  3
   1 27  6  4
   2  0 16  0
   3  0  5 21
```

There are 79 objects, and hence 3081 different pairs of objects. So for the comparison of interest we need to find out how many of those pairs went in the same cluster in both samples, how many pairs went in different clusters (these are both concordant values), and finally how many pairs were in one cluster, under one algorithm, but not in one cluster in the other.

```
>  outSamekm3 = outer(km3$cluster, km3$cluster,
                      FUN = function(x,y) x==y)
>  outSamepam3 = outer(pam3$clustering, pam3$clustering,
                       FUN = function(x,y) x==y)
>  inSBoth = outSamekm3 & outSamepam3
>  ##then we subtract 79, because an obs is in the same
>  ## cluster as itself this just means that the diagonal
>  ## is TRUE and divide by two
>  sameBoth = (sum(inSBoth) - 79)/2
>  ##not in the same one, in both are those FALSE entries
>  notSBoth = (!outSamekm3) & (!outSamepam3)
>  notSameBoth = sum(notSBoth)/2
>  ##those that are different, are TRUE in one and FALSE
>  ## in the other or vice versa
>
>  diffBoth = ((!outSamekm3) & outSamepam3) |
        (outSamekm3 & (!outSamepam3))
>  discordant = sum(diffBoth)/2
```

Thus we see that there are 712 pairs that are put in the same group in both clusterings. There are 1680 pairs that were not put in the same cluster, for either algorithm. And there were 689 pairs that were put in the same cluster for one of the algorithms, but not for the other.

Exercise 10.8

Basically repeat the steps given above, only now the table will have three groups instead of two.

Exercise 10.9

The first run has the samples more evenly spread among the 16 groups; there are no clusters of size 1, whereas for both other methods there are a number of clusters of size 1. In the code below we first show the distribution of samples among the 16 clusters for method 1, and then the number of clusters of different sizes for method 2. Notice that most of the clusters are of size 1.

```
> table(s1$unit.classif)
 1  2  3  4  5  6  7  8  9 10 11 12 13 14 15 16
 8  3  9  5  5  3  3  5  3  2  5  5  8  5  4  6
> table(table(s2$unit.classif))
 1  2  7  9 16 19
10  1  1  1  2  1
> table(table(s3$unit.classif))
 1  2  7  8  9 10 15 19
 9  1  1  1  1  1  1  1
```

   Comparing clusterings is a bit hard, as there are no obvious labels to put on the clusters. We might want to devise some code that will tell us which samples were clustered together in two methods, and which number were together in one, not together in the other, and finally how many pairs were in different clusters for both outputs. But this is another topic, and one for which we do not have room.

Exercise 10.10

We first identify the samples, and then subset the expression values. Then we define some colors, so that we can easily tell the two groups apart.

```
> intOnes = s1$unit.classif == 13 | s1$unit.classif == 14
> gvsub = gvals[intOnes,]
> gvclasses = s1$unit.classif[intOnes]
> sideC = ifelse(gvclasses==13, "yellow", "blue")
> heatmap(t(gvsub), ColSideCol=sideC)
```

Exercise 10.11

So, we want to repeat our *k*-means analysis, as described above but with $k = 4$.

```
> km2sol = kmeans(gvals, centers=4, nstart=25)
> table(km2sol$cluster, SOMgp2)
```

```
    SOMgp2
        1  2  3  4  5  6  7
    1   0 12  1  2  1  0  5
    2   3  5  2  0  0 12  0
    3   0  0  1  0  0  2 13
    4   0  3  7  2  4  3  1
```

The results seem remarkably concordant. Next we remove those samples that correspond to the smaller groups and repeat the k-means analysis.

```
> dropInds = SOMgp2 %in% c("1", "4", "5")
> gvals2 = gvals[!dropInds,]
> km3 = kmeans(gvals2, centers=4, nstart=50)
> table(km3$cluster, SOMgp2[!dropInds])
        1  2  3  4  5  6  7
    1   0 12  1  0  0  0  6
    2   0  1  0  0  0 14  0
    3   0  0  1  0  0  2 13
    4   0  7  9  0  0  1  0
```

Exercise 10.12

This can be found directly from `silpam2` by mimicking the code above.

```
>    ans = silpam2[silpam2[, "sil_width"] < 0, ]
```

So there are five observations with negative silhouette widths.

Exercise 10.13

For this problem we make use of the output of `diana`, and hence work with hc4.

```
>    dcl4 = cutree(hc4, 4)
>    table(dcl4)
dcl4
  1  2  3  4
 28 15 22 14
>  ## we presume the labels are in the order
>  ## given to the \indexTerm{clustering} algorithm
>
>  sild4 =  silhouette(dcl4, manDist)
```

We can compute the silhouette, and using the methods discussed in Section 10.8 we can plot this, or perform other operations on it.

Exercise 10.14

An example probe set is 33232_at. Try

```
> t.test(exprs(esTT)["33232_at",]~esTT$mol.biol)
     Welch Two Sample t-test

data:  exprs(esTT)["33232_at", ] by esTT$mol.biol
t = 4.46, df = 76.8, p-value = 2.726e-05
alternative hypothesis: true difference in means is not
equal to 0
95 percent confidence interval:
 0.71 1.85
sample estimates:
mean in group BCR/ABL     mean in group NEG
                8.66                   7.37
```

Exercise 10.15

One way to do this is to increase the value of K in:

```
> esTT.K = ALLfilt_bcrneg[ordtt[1:K],]
```

and then repeat the steps shown above to develop the principal components and biplot visualizations.

11 Using Graphs for Interactome Data

Exercise 11.1

   a RobjectlitG is an instance of the *graphNEL* class. You can use the manual page to find out more; type class?graphNEL.

   b
```
> nodes(litG)[1:5]
  [1] "YBL072C" "YBL083C" "YBR009C" "YBR010W" "YBR031W"
```

   c The ccyclered data is stored in a *data.frame*. Please also have a look at the manual page of this class.

```
> class(ccyclered)
> str(ccyclered)
> dim(ccyclered)
> names(ccyclered)
```

Exercise 11.2

a Each element of `cc` is a character vector of node names defining one of the connected components of the graph.

```
> cc[[7]]
[1] "YBR118W" "YAL003W" "YLR249W"
```

b There are 2642 connected components. The largest component consists of 88 nodes. There are 2587 singletons.

Exercise 11.3

Again, we first create the layouts using the function `layoutGraph`. These can then be plotted using `renderGraph`, as above.

```
> lay12neato = layoutGraph(sg1, layoutType="dot")
> renderGraph(lay12neato,
       graph.pars=list(nodes=list(fillcolor="steelblue2")))
> lay12twopi = layoutGraph(sg2, layoutType="twopi")
> renderGraph(lay12twopi,
       graph.pars=list(nodes=list(fillcolor="steelblue2")))
```

Exercise 11.4

The `sps` object is a list. The manual page for `sp.between` describes its structure.

To plot individual nodes and edges with different colors we have to use the `nodeRenderInfo` and `edgeRenderInfo` functions:

```
> fill = rep("steelblue2", length(pth))
> names(fill) = pth
> nodeRenderInfo(lsg1) = list(fill=fill)
> edges = paste(pth[-length(pth)], pth[-1], sep="~")
> lwd = rep(5, length(edges))
> col = rep("steelblue2", length(edges))
> names(lwd) = names(col) = edges
> edgeRenderInfo(lsg1) = list(col=col, lwd=lwd)

> renderGraph(lsg1)
```

Exercise 11.5

`allp` is a matrix of the shortest path distances between all pairs of nodes. You can find the diameter by finding the maximum value in `allp`.

```
> max(allp)
[1] 13
```

The longest shortest path is not unique:

```
> sum(allp == max(allp))
[1] 36
```

Exercise 11.6

```
> clusts = with(ccyclered, split(Y.name, Cluster))
```

Exercise 11.7

```
> ccClust = connectedComp(cg)
> length(ccClust)
[1] 30
```

Exercise 11.8

The return value of the `intersection` method is a new *graph* object containing the common set of nodes and edges between the two input graphs. So the number of common edges between the graphs is simply the number of edges in the returned graph:

```
> numEdges(commonG)
[1] 42
```

Exercise 11.9

The `nodePerm` function takes two graphs, g1 and g2, as inputs along with the number of permutation-based tests to perform, B. The function loops B times. For each iteration, the node labels of g1 are permuted and the number of common edges between the permuted g1 and g2 is computed. The return value is a numeric vector with length equal to B such that each element gives the number of common edges for the corresponding permutation.

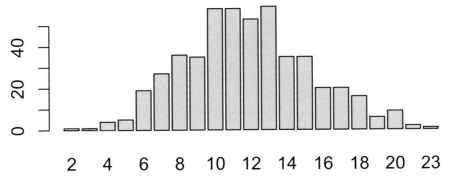

Figure 15.4. Barplot of frequencies in the permutation distribution of the number of common edges in `nPdist`. The observed number of common edges `numEdges(commonG)=42` is larger than any of the permutation values, which indicates that there are significantly more common edges than what would be expected if co-expression and protein interaction were unrelated.

Exercise 11.10

```
> barplot(table(nPdist))
```

12   Graph Layout

Exercise 12.1

```
> graph.par(list(graph=list("cex.main"=2.5)))
> x = layoutGraph(g, layoutType="neato")
> renderGraph(x, graph.pars=list(graph=list(main="neato")))


> x = layoutGraph(g, layoutType="twopi")
> renderGraph(x, graph.pars=list(graph=list(main="twopi")))


> x = layoutGraph(g, layoutType="circo")
> renderGraph(x, graph.pars=list(graph=list(main="circo")))


> x = layoutGraph(g, layoutType="fdp")
> renderGraph(x, graph.pars=list(graph=list(main="fdp")))
```

Exercise 12.2

This exercise is open-ended and it has no unique solution. Consult the
documentation of `renderParameters` for its parameters.

```
> ? renderParameters
```

Exercise 12.3

Currently **Rgraphviz** supports the following node shapes: "circle",
"ellipse", and "rect".

```
> ? layoutParameters
```

`drawNodes` can be used for user-defined node plotting. It takes a function
which will be called to render a single node. Alternatively, a named list of
functions with list names equal to node names can be used to render each
node differently.

Exercise 12.4

```
> colors = rep("lightgreen", length(nodes(IMCAGraph)))
> names(colors) = nodes(IMCAGraph)
> transp = c("ITGB", "ITGA", "MYO", "ACTN", "JNK", "p110",
      "Phosphatidylinositol signaling system",
      "PI5K", "MYO-P", "cell maintenance", "cell motility",
      "F-actin", "cell proliferation")
> colors[transp] = "transparent"
> nodeRenderInfo(IMCAGraph) = list(fill=colors)
> renderGraph(IMCAGraph)
```

Exercise 12.5

```
> sg4 = subGraph(c("GRB2", "SOS", "Ha-Ras", "Raf",
      "MEK", "ERK"), IMCAGraph)
> subGList = append(subGList, list(list(graph=sg4)))
> IMCAGraph = layoutGraph(IMCAGraph, attrs=attrs,
      nodeAttrs=nodeAttrs, subGList=subGList)
> renderGraph(IMCAGraph)
```

13   Gene Set Enrichment Analysis

Exercise 13.1

  a There are 79 samples, and the table below shows how many of each type.

```
> table(ALLfilt_bcrneg$mol.biol)

BCR/ABL      NEG
    37       42
```

  b 5038 distinct genes (as determined by Entrez Gene ID) have been selected.

Exercise 13.2

There is one gene set for each row of the incidence matrix, so there are 194 gene sets. We can find out how many gene sets have fewer than 10 genes by computing `rowSums(Am)`, so there are 67 gene sets. The largest number of gene sets a gene is in can be found by finding the largest of the column sums, which is 32 gene sets for `976_s_at`.

Exercise 13.3

There are 794 positive statistics, and 872 negative ones. There are 134 with p-values less than 0.01.

Exercise 13.4

You should notice that all of the points lie above the 45 degree line, indicating that they have higher y values than x values. Or, that the mean value in the NEG group is larger than the mean value in the BCR/ABL group.

Exercise 13.5

Although the mean plot seemed to suggest a strong separation between the two groups, we see from the heatmap that the distinction is not that clear.

    The row in the heatmap that corresponds to the gene labeled `41214_at` indicates that the gene is on in some samples and off in others. It is on the Y chromosome, and hence we are seeing a pattern of expression that distinguishes the male samples from the females.

Exercise 13.6

```
> apply(pvals, 2, min)
Lower Upper
0.022 0.000
> rownames(pvals)[apply(pvals, 2, which.min)]
[1] "03010" "04510"
```

Exercise 13.7

To obtain the *p*-values from the permutation approach we must obtain the rowwise minima of the permutation *p*-values. The *p*-values for the parametric approach can be obtained by calling pnorm, and then taking the smaller of the observed *p*-value or one minus it.

```
> permpvs  = pmin(pvals[,1], pvals[,2])
> pvsparam = pnorm(tAadj)
> pvspara  = pmin(pvsparam, 1-pvsparam)


> plot(permpvs, pvspara, xlab="Permutation p-values",
        ylab="Parametric p-values")
```

Exercise 13.8

It indicates that the gene is on the p arm of chromosome 17 in band 3, subband 3, subsubband 2.

Exercise 13.9

```
> ## depending on which annotation infrastructure we use
> ## hgu95av2MAP will either be an environment or an
> ## AnnDbBimap object
> fnames = featureNames(ALLfilt_bcrneg)
> if(is(hgu95av2MAP, "environment")){
      chrLocs = mget(fnames, hgu95av2MAP)
      mapping = names(chrLocs[sapply(chrLocs,
          function(x) !all(is.na(x)))])
  }else{
      mapping = toTable(hgu95av2MAP[fnames])$probe_id
  }
> psWithMAP = unique(mapping)
> nsF2 = ALLfilt_bcrneg[psWithMAP, ]
```

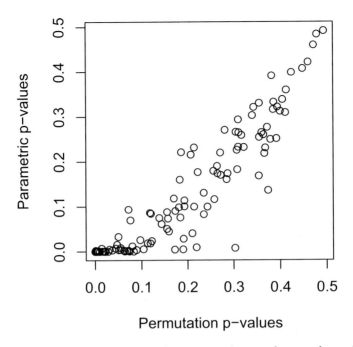

Figure 15.5. Scatter plot comparing the permutation p-values to those obtained from using a Normal approximation.

Exercise 13.10

The value returned by MAPAmat is a matrix where the rows are the chromosome bands and the columns are the genes. So there are 5003 genes and 1070 map positions.

```
> dim(chrMat)
[1] 1070 5003
```

Exercise 13.11

```
> chrMat = chrMat[rowSums(chrMat) >= 5, ]
> dim(chrMat)
[1]  547 5003
```

Exercise 13.12

```
> EGlist = mget(featureNames(nsF2),  hgu95av2ENTREZID)
> EGIDs = sapply(EGlist, "[", 1)
```

```
> idx = match(EGIDs, colnames(chrMat))
> chrMat = chrMat[, idx]
```

Now you can simply repeat the analysis using the new incidence matrix chrMat.

Exercise 13.13

```
> rowSums(Ams)[c("04510", "04512", "04514", "04940")]
04510 04512 04514 04940
   96    31    52    27
> Amx["04512", "04510"]
[1] 21
> Amx["04940", "04514"]
[1] 17
```

Exercise 13.14

Example code

```
> P04514 = Ams["04514",]
> P04940 = Ams["04940",]
> P04514.Only = ifelse(P04514 != 0 & P04940 == 0, 1, 0)
> P04940.Only = ifelse(P04514 == 0 & P04940 != 0, 1, 0)
> Both        = ifelse(P04514 != 0 & P04940 != 0, 1, 0)
> lm5 = lm(rttStat ~ P04514.Only + P04940.Only + Both)
> summary(lm5)
Call:
lm(formula = rttStat ~ P04514.Only + P04940.Only + Both)

Residuals:
   Min    1Q Median    3Q    Max
-4.173 -1.052 -0.169  0.865  7.205

Coefficients:
            Estimate Std. Error t value Pr(>|t|)
(Intercept)   0.0751     0.0381    1.97   0.0486 *
P04514.Only   0.7490     0.2605    2.87   0.0041 **
P04940.Only   0.7346     0.4837    1.52   0.1290
Both          1.0056     0.3718    2.70   0.0069 **
---
Signif. codes:  0 '***' 0.001 '**' 0.01 '*' 0.05 '.' 0.1 ' ' 1

Residual standard error: 1.52 on 1662 degrees of freedom
```

```
Multiple R-Squared: 0.0104, Adjusted R-squared: 0.00863
F-statistic: 5.83 on 3 and 1662 DF,  p-value: 0.000581
```

The answer is a bit less clear here. Genes only in 04514 give an extremely small *p*-value, whereas those in both and those only in 04940 have a lesser effect.

14   Hypergeometric Testing Used for Gene Set Enrichment Analysis

Exercise 14.1

a
```
> numSamp = length(ALL_bcrneg$mol.biol)
> table(ALL_bcrneg$mol.biol)
```
```
BCR/ABL      NEG
     37       42
```

b
```
> annotation(ALL_bcrneg)
[1] "hgu95av2"
> length(featureNames(ALL_bcrneg))
[1] 12625
```

Exercise 14.2

See the description of the `remove.dupEntrez` argument in the manual page for `nsFilter`.

Exercise 14.3

The `filter.log` component of the `nsFilter` return value provides information about the number of probe sets removed by each step of the filter.

Exercise 14.4

```
> chrN = mget(featureNames(ALLfilt_bcrneg), envir=hgu95av2CHR)
> onY = sapply(chrN, function(x) any(x == "Y"))
> onY[is.na(onY)] = FALSE
> ALLfilt_bcrneg = ALLfilt_bcrneg[!onY, ]
```

Exercise 14.5

```
> ## an alternate universe based on the entire chip
> chipAffyUniverse = featureNames(ALLfilt_bcrneg)
> chipEntrezUniverse = mget(chipAffyUniverse, hgu95av2ENTREZID)
> chipEntrezUniverse = unique(unlist(chipEntrezUniverse))
```

Exercise 14.6

```
> sumpv = sum(smPV)
```

There are 713 probe sets with p-values less than 0.05.

Exercise 14.7

```
a > df = summary(hgOver)
  > names(df)

  [1] "GOBPID"    "Pvalue"    "OddsRatio" "ExpCount"
  [5] "Count"     "Size"      "Term"
```

```
b > df = summary(hgOver, pvalue=0.05, categorySize=350)
  > nrow(df)

  [1] 21
```

```
c > ? HyperGResult-accessors
```

Exercise 14.8

```
> browseURL("ALL_hgo.html")
```

Exercise 14.9

```
> numG = length(sigSub)
> sizes = sapply(sigSub, numNodes)
> sizes
1 2 3
3 2 1
```

are displayed above.

Exercise 14.10

a
```
> dfcond = summary(hgCond, categorySize=50)
> ## trim the term names for display purposes
> trimTerm = function(x) {
      if (nchar(x) <= 20)
          x
      else
          paste(substr(x, 1, 20), "...", sep="")
  }
> dfcond$Term = sapply(dfcond$Term, trimTerm)
> sizeOrd = order(dfcond$Size, decreasing=TRUE)
> dfcond[sizeOrd, c("Count", "Size", "Term")]
           Count Size              Term
GO:0007154   241 1170   cell communication
GO:0006955    65  257      immune response
GO:0007155    57  209         cell adhesion
GO:0009966    51  182 regulation of signal...
```

b
```
> stdIds = sigCategories(hgOver)
> condIds = sigCategories(hgCond)
> setdiff(stdIds, condIds)

[1] "GO:0007165" "GO:0022610"
```

Exercise 14.11

```
> params = new("ChrMapHyperGParams",
      conditional=FALSE, testDirection="over",
      universeGeneIds=entrezUniverse,
      geneIds=selectedEntrezIds,
      annotation="hgu95av2", pvalueCutoff=0.05)
> paramsCond = params
> conditional(paramsCond) = TRUE

> hgans = hyperGTest(params)
> hgansCond = hyperGTest(paramsCond)

> summary(hgans, categorySize=10)
     ChrMapID  Pvalue OddsRatio ExpCount Count Size
7p15     7p15 0.000644      4.98     3.38    10   20
1q21     1q21 0.000823      2.93     7.77    17   46
7p         7p 0.008175      2.04    11.65    20   69
```

8q24	8q24 0.009560	2.73	5.23	11	31
7p1	7p1 0.012335	2.20	8.27	15	49
3q25	3q25 0.013347	4.25	2.19	6	13
14q22	14q22 0.039695	2.97	2.70	6	16
7	7 0.046208	1.38	32.75	42	194

Exercise 14.12

```
> kparams = new("KEGGHyperGParams",
    geneIds=selectedEntrezIds,
    universeGeneIds=entrezUniverse,
    annotation="hgu95av2",
    pvalueCutoff=0.05,
    testDirection="over")
> kans = hyperGTest(kparams)

> summary(kans)
      KEGGID Pvalue OddsRatio ExpCount Count Size
04916  04916 0.0215     2.18     7.32    13   42
04810  04810 0.0220     1.68    17.78    26  102
05217  05217 0.0230     3.61     2.44     6   14
04640  04640 0.0290     2.00     8.37    14   48
04520  04520 0.0339     2.08     6.97    12   40
04360  04360 0.0370     1.87     9.41    15   54
04510  04510 0.0473     1.58    16.38    23   94
                                    Term
04916                       Melanogenesis
04810 Regulation of actin cytoskeleton
05217                Basal cell carcinoma
04640          Hematopoietic cell lineage
04520                   Adherens junction
04360                        Axon guidance
04510                      Focal adhesion
> kparamsUnder = kparams
> testDirection(kparamsUnder) = "under"

> kansUnder = hyperGTest(kparamsUnder)

> summary(kansUnder)
      KEGGID Pvalue OddsRatio ExpCount Count Size
04650  04650 0.0100    0.353    12.20     5   70
04664  04664 0.0229    0.251     6.80     2   39
05211  05211 0.0229    0.251     6.80     2   39
```

```
00020    00020  0.0257        0.000        3.31      0    19
00380    00380  0.0369        0.179        4.71      1    27
00051    00051  0.0379        0.000        2.96      0    17
04080    04080  0.0412        0.394        8.89      4    51
04012    04012  0.0412        0.394        8.89      4    51
04370    04370  0.0412        0.282        6.10      2    35
00071    00071  0.0434        0.187        4.53      1    26
                                                    Term
04650  Natural killer cell mediated cytotoxicity
04664                 Fc epsilon RI signaling pathway
05211                       Renal cell carcinoma
00020                    Citrate cycle (TCA cycle)
00380                       Tryptophan metabolism
00051              Fructose and mannose metabolism
04080    Neuroactive ligand-receptor interaction
04012                        ErbB signaling pathway
04370                        VEGF signaling pathway
00071                         Fatty acid metabolism
```

Exercise 14.13

```
> pparams = new("PFAMHyperGParams",
     geneIds=selectedEntrezIds,
     universeGeneIds=entrezUniverse,
     annotation="hgu95av2",
     pvalueCutoff=hgCutoff,
     testDirection="over")
> pans = hyperGTest(pparams)

> summary(pans)
          PFAMID  Pvalue OddsRatio ExpCount Count Size
PF01023 PF01023 0.000129     30.12    1.171     6    7
PF01833 PF01833 0.000365      6.47    2.676     9   16
PF07714 PF07714 0.000402      3.19    7.360    17   44
PF08337 PF08337 0.000777       Inf    0.669     4    4
            Term
PF01023 PF01023
PF01833 PF01833
PF07714 PF07714
PF08337 PF08337
```

References

A. Alexa, J. Rahnenfuhrer, and T. Lengauer. Improved scoring of functional groups from gene expression data by decorrelating GO graph structure. *Bioinformatics*, 22(13):1600–7, 2006.

R. Balasubramanian, T. LaFramboise, D. Scholtens, et al. A graph theoretic approach to testing associations between disparate sources of functional genomics data. *Bioinformatics*, 20:3353–3362, 2004.

Y. Benjamini and Y. Hochberg. Controlling the false discovery rate: A practical and powerful approach to multiple testing. *JRSSB*, 57:289–300, 1995.

B. M. Bolstad, R. A. Irizarry, L. Gautier, and Z. Wu. Preprocessing high-density oligonucleotide arrays. In R. Gentleman, W. Huber, V. Carey, R. Irizarry, and S. Dudoit, editors, *Bioinformatics and Computational Biology Solutions Using R and Bioconductor*. Springer, New York, 2005.

L. Breiman. Random forests – random features. *Preprint Department of Statistics U.C. Berkeley*, (567), 1999.

L. Breiman, J. H. Friedman, R. A. Olshen, et al. *Classification and Regression Trees*. Wadsworth, Belmont, CA, 1984.

J. Brettschneider, F. Collin, B. Bolstad, and T. Speed. Quality assessment for short oligonucleotide arrays. *Technometrics, in press*, 2007.

V. Carey, J. Gentry, E. Whalen, and R. Gentleman. Network structures and algorithms in Bioconductor. *Bioinformatics*, 21:135–136, 2005.

S. Chiaretti, X. Li, R. Gentleman, A. Vitale, M. Vignetti, F. Mandelli, J. Ritz, and R. Foa. Gene expression profile of adult T-cell acute lymphocytic leukemia identifies distinct subsets of patients with different response to therapy and survival. *Blood*, 103:2771–2778, 2004.

S. Chiaretti, X. Li, R. Gentleman, A. Vitale, K. S. Wang, F. Mandelli, R. Foa, and J. Ritz. Gene expression profiles of b-lineage adult acute lymphocytic leukemia reveal genetic patterns that identify lineage derivation and distinct mechanisms of transformation. *Clinical Cancer Research*, 11:7209–7219, 2005.

R. Cho, M. Campbell, E. Winzeler, et al. A genome-wide transcriptional analysis of the mitotic cell cycle. *Molecular Cell*, 2:65–73, 1998.

S. Dudoit, Y. H. Yang, T. P. Speed, and M. J. Callow. Statistical methods for identifying differentially expressed genes in replicated cDNA microarray experiments. *Statistica Sinica*, 12:111–139, 2002.

B. P. Durbin, J. S. Hardin, D. M. Hawkins, and D. M. Rocke. A variance-stabilizing transformation for gene-expression microarray data. *Bioinformatics*, 18 Suppl 1:105–110, 2002.

ENCODE Project Consortium, E. Birney, J. A. Stamatoyannopoulos, A. Dutta, et al. Identification and analysis of functional elements in 1% of the human genome by the ENCODE pilot project. *Nature*, 447:799–816, 2007.

S. Falcon and R. Gentleman. Using GOstats to test gene lists for GO term association. *Bioinformatics*, 23(2):257–8, 2007.

R. D. Finn, J. Mistry, B. Schuster-Bockler, S. Griffiths-Jones, V. Hollich, T. Lassmann, S. Moxon, M. Marshall, A. Khanna, R. Durbin, S. R. Eddy, E. L. L. Sonnhammer, and A. Bateman. Pfam: Clans, Web tools and services. *Nucleic Acids Res*, 34(Database issue):247–251, Jan 2006.

P. Flicek, B. L. Aken, K. Beal, B. Ballester, M. Caccamo, Y. Chen, L. Clarke, G. Coates, F. Cunningham, T. Cutts, T. Down, S. C. Dyer, T. Eyre, S. Fitzgerald, J. Fernandez-Banet, S. Gräf, S. Haider, M. Hammond, R. Holland, K. L. Howe, K. Howe, N. Johnson, A. Jenkinson, A. Kähäri, D. Keefe, F. Kokocinski, E. Kulesha, D. Lawson, I. Longden, K. Megy, P. Meidl, B. Overduin, A. Parker, B. Pritchard, A. Prlic, S. Rice, D. Rios, M. Schuster, I. Sealy, G. Slater, D. Smedley, G. Spudich, S. Trevanion, A. J. Vilella, J. Vogel, S. White, M. Wood, E. Birney, T. Cox, V. Curwen, R. Durbin, X. M. Fernandez-Suarez, J. Herrero, T. J. Hubbard, A. Kasprzyk, G. Proctor, J. Smith, A. Ureta-Vidal, and S. Searle. Ensembl 2008. *Nucleic Acids Res*, 2007.

E. R. Gansner and S. C. North. An open graph visualization system and its applications to software engineering. *Software Practice and Experience*, 30:1203–1233, 1999.

H. Ge, Z. Liu, G. M. Church, and M. Vidal. Correlation between transcriptome and interactome mapping data from Saccharomyces cerevisiae. *Nature Genetics*, 29:482–486, 2001.

R. Gentleman, W. Huber, V. Carey, R. Irizarry, and S. Dudoit, editors. *Bioinformatics and Computational Biology Solutions Using R and Bioconductor*. Springer, New York, 2005a.

R. Gentleman, M. Ruschhaupt, and W. Huber. On the synthesis of microarray experiments. *Journal de la Société Françaises de Statistique*, 146 (1-2):173–194, 2005b.

R. C. Gentleman, V. J. Carey, D. M. Bates, et al. Bioconductor: Open software development for computational biology and bioinformatics. *Genome Biology*, 5: R80, 2004.

A. D. Gordon. *Classification*. Chapman & Hall CRC, Boca Raton, FL, 2nd edition, 1999.

T. Hastie, R. Tibshirani, and J. H. Friedman. *The Elements of Statistical Learning: Data Mining, Inference, and Prediction*. Springer-Verlag, New York, 2001.

W. Huber, A. von Heydebreck, H. Sültmann, A. Poustka, and M. Vingron. Variance stablization applied to microarray data calibration and to quantification of differential expression. *Bioinformatics*, 18:96–104, 2002.

W. Huber, A. von Heydebreck, H. Sültmann, A. Poustka, and M. Vingron. Parameter estimation for the calibration and variance stabilization of microarray data. *Statistical Applications in Genetics and Molecular Biology*, 2(1), 2003.

W. Huber, A. von Heydebreck, and M. Vingron. *Encyclopedia of Genetics, Genomics, Proteomics and Bioinformatics*, chapter Error models for microarray intensities. Wiley, New York, 2005.

R. A. Irizarry, B. Hobbs, F. Collin, Y. D. Beazer-Barclay, K. J. Antonellis, U. Scherf, and T. P. Speed. Exploration, normalization, and summaries of high density oligonucleotide array probe level data. *Biostatistics*, 4:249–264, 2003.

Z. Jiang and R. Gentleman. Extensions to gene set enrichment. *Bioinformatics*, 23:306–313, 2007.

M. Kanehisa and S. Goto. KEGG: Kyoto encyclopedia of genes and genomes. *Nucleic Acids Res*, 28:27–30, 2000.

M. Kanehisa, S. Goto, M. Hattori, K. F. Aoki-Kinoshita, M. Itoh, S. Kawashima, T. Katayama, M. Araki, and M. Hirakawa. From genomics to chemical genomics: new developments in KEGG. *Nucleic Acids Res*, 34(Database issue):354–357, Jan 2006.

L. Kaufman and P. J. Rousseeuw. *Finding Groups in Data*. Wiley, New York, 1990.

T. Kohonen. *Self Organizing Maps*. Springer, Berlin, 1995.

S. K. Kummerfeld and S. A. Teichmann. DBD: A transcription factor prediction database. *Nucleic Acids Res*, 34:D74–D81, 2006.

A. Liaw and M. Wiener. Classification and regression by randomforest. *R News*, 2(3):18–22, 2002. URL http://CRAN.R-project.org/doc/Rnews.

I. Lönnstedt and T. Speed. Replicated microarray data. *Statistica Sinica*, 12: 31–46, 2002.

D. Maglott, J. Ostell, K. D. Pruitt, and T. Tatusova. Entrez Gene: gene-centered information at NCBI. *Nucleic Acids Res*, 35(Database issue):26–31, 2007.

M. McGee and Z. Chen. Parameter estimation for the exponential-normal convolution model for background correction of Affymetrix GeneChip data. *Statistical Applications in Genetics and Molecular Biology*, 5(1):Article 24, 2006.

N. J. Mulder, R. Apweiler, T. K. Attwood, A. Bairoch, A. Bateman, D. Binns, P. Bork, V. Buillard, L. Cerutti, R. Copley, E. Courcelle, U. Das, L. Daugherty, M. Dibley, R. Finn, W. Fleischmann, J. Gough, D. Haft, N. Hulo, S. Hunter, D. Kahn, A. Kanapin, A. Kejariwal, A. Labarga, P. S. Langendijk-Genevaux, D. Lonsdale, R. Lopez, I. Letunic, M. Madera, J. Maslen, C. McAnulla, J. McDowall, J. Mistry, A. Mitchell, A. N. Nikolskaya, S. Orchard, C. Orengo, R. Petryszak, J. D. Selengut, C. J. Sigrist, P. D. Thomas, F. Valentin, D. Wilson,

C. H. Wu, and C. Yeats. New developments in the interpro database. *Nucleic Acids Res*, 35(Database issue):224–228, 2007.

P. Murrell. *R Graphics*. Chapman and Hall, Boca Raton, FL, 2005.

M. S. Pepe, G. Longton, G. L. Anderson, and M. Schummer. Selecting differentially expressed genes from microarray experiments. *Biometrics*, 59(1):133–142, Mar 2003.

K. S. Pollard and M. J. van der Laan. Cluster analysis of genomic data. In R. Gentleman, W. Huber, V. Carey, R. Irizarry, and S. Dudoit, editors, *Bioinformatics and Computational Biology Solutions Using R and Bioconductor*. Springer, New York, 2005.

K. S. Pollard, S. Dudoit, and M. J. van der Laan. Multiple testing procedures: The multtest package and applications to genomics. In R. Gentleman, W. Huber, V. Carey, R. Irizarry, and S. Dudoit, editors, *Bioinformatics and Computational Biology Solutions Using R and Bioconductor*. Springer, New York, 2005.

R-Foundation. Introduction to R, 2007. URL http://cran.r-project.org/manuals/R-intro.html.

B. D. Ripley. *Pattern Recognition and Neural Networks*. Cambridge University Press, Cambridge, 1996.

M. E. Ritchie, J. Silver, A. Oshlack, M. Holmes, D. Diyagama, A. Holloway, and G. K. Smyth. A comparison of background correction methods for two-colour microarrays. *Bioinformatics*, 23:2700–2707, 2007.

D. M. Rocke and B. Durbin. Approximate variance-stabilizing transformations for gene-expression microarray data. *Bioinformatics*, 19:966–972, 2003.

P. J. Rousseuw and A. M. Leroy. *Robust Regression and Outlier Detection*. Wiley, New York, 1987.

A. Schroeder, O. Mueller, S. Stocker, R. Salowsky, M. Leiber, M. Gassmann, S. Lightfoot, W. Menzel, M. Granzow, and T. Ragg. The RIN: an RNA integrity number for assigning integrity values to RNA measurements. *BMC Mol Biol*, 7: 3, 2006.

J. G. Siek, L.-Q. Lee, and A. Lumsdaine. *The Boost Graph Library*. Addison Wesley, Boston, 2002.

G. Smyth. Linear models and empirical Bayes methods for assessing differential expression in microarray experiments. *Statistical Applications in Genetics and Molecular Biology*, 3:Article 3, 2004.

P. H. A. Sneath and R. R. Sokal. *Numerical Taxonomy: The Principles and Practice of Numerical Classification*. Freeman, San Francisco, 1973.

A. Subramanian, P. Tamayo, V. K. Mootha, et al. Gene set enrichment analysis: A knowledge-based approach for interpreting genome-wide expression profiles. *Proc. Natl. Acad. Sci. of the U.S.A.*, 102(43):15545–15550, 2005.

H. Sültmann, A. von Heydebreck, W. Huber, R. Kuner, A. Buness, M. Vogt, B. Gunawan, M. Vingron, L. Füzesí, and A. Poustka. Gene expression in kidney cancer is associated with cytogenetic abnormalities, metastasis formation, and patient survival. *Clinical Cancer Research*, 11:646–655, 2005.

The Gene Ontology Consortium. Gene Ontology: tool for the unification of biology. *Nature Genetics*, 25:25–29, 2000.

L. Tian, S. A. Greenberg, S. W. Kong, et al. Discovering statistically significant pathways in expression profiling studies. *Proc. Natl. Acad. Sci. of the U.S.A.*, 102(38):13544–13549, 2005.

V. Tusher, R. Tibshirani, and G. Chu. Significance analysis of microarrays applied to the ionizing radiation response. *Proc. Natl. Acad. Sci. of the U.S.A.*, 98: 5116–5121, 2001.

UniProt. The universal protein resource (UniProt). *Nucleic Acids Res*, 35 (Database issue):193–197, 2007.

W. N. Venables and B. D. Ripley. *Modern Applied Statistics with S (4e)*. Springer, New York, 2002.

A. von Heydebreck, W. Huber, and R. Gentleman. Differential expression with the Bioconductor project. In *Encyclopedia of Genetics, Genomics, Proteomics and Bioinformatics*. Wiley, New York, 2004.

P. Westfall and S. Young. *Resampling-Based Multiple Testing: Examples and Methods for p-Value Adjustment*. Wiley, New York, 1993.

D. L. Wheeler, T. Barrett, D. A. Benson, S. H. Bryant, K. Canese, V. Chetvernin, D. M. Church, M. DiCuccio, R. Edgar, S. Federhen, L. Y. Geer, Y. Kapustin, O. Khovayko, D. Landsman, D. J. Lipman, T. L. Madden, D. R. Maglott, J. Ostell, V. Miller, K. D. Pruitt, G. D. Schuler, E. Sequeira, S. T. Sherry, K. Sirotkin, A. Souvorov, G. Starchenko, R. L. Tatusov, T. A. Tatusova, L. Wagner, and E. Yaschenko. Database resources of the National Center for Biotechnology Information. *Nucleic Acids Res*, 35(Database issue):5–12, 2007.

Index

 **Springer**
the language of science

springer.com

Bioinformatics and Computational Biology Solutions Using R and Bioconductor

R. Gentleman, V. Carey, W. Huber, R. Irizarry and S. Dudoit. (Eds.)

Bioconductor is a widely used open source and open development software project for the analysis and comprehension of data arising from high-throughput experimentation in genomics and molecular biology. Bioconductor is rooted in the open source statistical computing environment R. This volume's coverage is broad and ranges across most of the key capabilities of the Bioconductor project, including importation and preprocessing of high-throughput data from microarray, proteomic, and flow cytometry platforms.

2005. 473 pp. (Statistics for Biology and Health) Hardcover
ISBN 978-0-387-25146-2

Lattice: Multivariate Data Visualizations with R

Deepayan Sarkar

The book contains close to150 figures produced with lattice. Many of the examples emphasize principles of good graphical design; almost all use real data sets that are publicly available in various R packages. All code and figures in the book are also available online, along with supplementary material covering more advanced topics.

2008. Approx. 290 pp. (Use R!) Softcover
ISBN 978-0-387-75968-5

Data Manipulation with R

Phil Spector

This book presents a wide array of methods applicable for reading data into R, and efficiently manipulating that data. In addition to the built-in functions, a number of readily available packages from CRAN (the Comprehensive R Archive Network) are also covered. All of the methods presented take advantage of the core features of R: vectorization, efficient use of subscripting, and the proper use of the varied functions in R that are provided for common data management tasks.

2008. 158 pp. (Use R!) Softcover
ISBN 978-0-387-74730-9

Easy Ways to Order▶   Call: Toll-Free 1-800-SPRINGER • E-mail: orders-ny@springer.com • Write: Springer, Dept. S8113, PO Box 2485, Secaucus, NJ 07096-2485 • Visit: Your local scientific bookstore or urge your librarian to order.

Printed in the United States